HITE 6.0 软件开发与应用工程师

工信部国家级计算机人才评定体系

使用 Android 高级技术开发 APP

武汉厚溥教育科技有限公司　编著

清华大学出版社
北京

内 容 简 介

本书按照高等院校、高职高专计算机课程基本要求，以案例驱动的形式来组织内容，突出计算机课程的实践性特点。本书共十三个单元，内容包括 Android 系统介绍、UI 布局详解、Activity 和 Intent、Android UI 布局详解、Handler 消息传递机制、UI 中级控件、Android 高级控件、Service 的生命周期、广播和通知、Android 中的数据存储、ContentProvider、网络编程和 XML/JSON 数据解析。

本书内容安排合理，层次清楚，通俗易懂，实例丰富，突出理论与实践的结合，可作为各类高等院校、高职高专及培训机构的教材，也可供广大 Android 应用开发人员参考。

本书封面贴有清华大学出版社防伪标签，无标签者不得销售。
版权所有，侵权必究。举报：010-62782989，beiqinquan@tup.tsinghua.edu.cn。

图书在版编目(CIP)数据

使用 Android 高级技术开发 APP / 武汉厚溥教育科技有限公司 编著. —北京：清华大学出版社，2020.8（2024.2重印）
(HITE 6.0 软件开发与应用工程师)
ISBN 978-7-302-54624-5

Ⅰ. ①使… Ⅱ. ①武… Ⅲ. ①移动终端－应用程序－程序设计 Ⅳ. ①TN929.53

中国版本图书馆 CIP 数据核字(2020)第 005386 号

责任编辑：刘金喜
封面设计：王　晨
版式设计：孔祥峰
责任校对：马遥遥
责任印制：杨　艳

出版发行：	清华大学出版社			
网　　址：	https://www.tup.com.cn, https://www.wqxuetang.com			
地　　址：	北京清华大学学研大厦 A 座	邮　　编：	100084	
社 总 机：	010-83470000	邮　　购：	010-62786544	
投稿与读者服务：	010-62776969, c-service@tup.tsinghua.edu.cn			
质 量 反 馈：	010-62772015, zhiliang@tup.tsinghua.edu.cn			
印 装 者：	涿州市鲅润文化传播有限公司			
经　　销：	全国新华书店			
开　　本：	185mm×260mm	印　张：19.5	字　数：462 千字	
版　　次：	2020 年 8 月第 1 版	印　次：2024 年 2 月第 2 次印刷		
定　　价：	79.00 元			

产品编号：084818-01

编委会

主 任：

 翁高飞　郭长庚

副主任：

 侯　枫　时生乐　席红旗　方党生

委　员：

 刘　烨　李阿红　屈　毅　赵小华
 李　焕　师　哲　魏　迎　张青青
 李　杰　刘世进　李　伟　唐永平

主　审：

 熊　勇　冯　玲

前言

　　Android 是一种基于 Linux 的自由及开放源代码的操作系统，主要应用于移动设备，如智能手机和平板电脑，由 Google 公司和开放手机联盟领导及开发。2005 年 8 月由 Google 收购注资。2007 年 11 月，Google 与 84 家硬件制造商、软件开发商及电信营运商组建开放手机联盟共同研发改良 Android 系统。第一部 Android 智能手机发布于 2008 年 10 月，随后，Android 逐渐扩展到平板电脑及其他领域，如电视、数码相机、游戏机等。2011 年第一季度，Android 在全球的市场份额首次超过 Symbian(塞班)系统，跃居全球第一。截至 2019 年 9 月，全世界采用这款系统的设备数量已超过 25 亿台。

　　本书是"工信部国家级计算机人才评定体系"中的一本专业教材。"工信部国家级计算机人才评定体系"是由武汉厚溥教育科技有限公司开发，以培养符合企业需求的软件工程师为目标的 IT 职业教育体系。在开发该体系之前，我们对 IT 行业的岗位序列做了充分的调研，包括研究从业人员技术方向、项目经验和职业素质等方面的需求，通过对所面向学生的特点、行业需求的现状及实施等方面详细分析，结合我公司对软件人才培养模式的认知，按照软件专业总体定位要求，进行软件专业产品课程体系设计。该体系集应用软件知识和多领域的实践项目于一体，着重培养学生的熟练度、规范性、集成和项目能力，从而达到预定的培养目标。

　　本书共十三个单元，内容包括 Android 系统介绍、UI 布局详解、Activity 和 Intent、Android UI 布局详解、Handler 消息传递机制、UI 中级控件、Android 高级控件、Service 的生命周期、广播和通知、Android 中的数据存储、ContentProvider、网络编程和 XML/JSON 数据解析。

　　我们对本书的编写体系做了精心的设计，按照"理论学习—知识总结—上机操作—课后习题"这一思路进行编排。"理论学习"部分描述通过案例要达到的学习目标与涉及的相关知识点，使学习目标更加明确；"知识总结"部分概括案例所涉及的知识点，使知识点完整系统地呈现；"上机操作"部分对案例进行了详尽分析，通过完整的步骤阐述帮助读者快速掌握该案例的操作方法；"课后习题"部分帮助读者理解章节的知识点。本书在内容编写方面，力求细致全面；在文字叙述方面，注意言简意赅、重点突出；在案例选取方面，强调案例的针对性和实用性。

本书凝聚了编者多年来的教学经验和成果，可作为各类高等院校、高职高专及培训机构的教材，也可供广大程序设计人员参考。

　　本书由武汉厚溥教育科技有限公司编著，由翁高飞、郭长庚、熊勇、冯玲等多名老师编写。本书编者长期从事项目开发和教学实施，并且对当前高校的教学情况非常熟悉，在编写过程中充分考虑不同学生的特点和需求，加强了项目实战方面的教学。在本书的编写过程中，得到了武汉厚溥教育科技有限公司各级领导的大力支持，在此对他们表示衷心的感谢。

　　参与本书编写的人员还有：许昌职业技术学院郭长庚，三门峡职业技术学院侯枫，咸阳职业技术学院李阿红、屈毅、赵小华、李焕、师哲、魏迎、张青青、李杰、刘世进，黄冈职业技术学院刘烨，梧州职业技术学院唐永平，河南财政金融学院席红旗，河南水利与环境职业学院时生乐，河南信息统计职业学院方党生，河南工业职业技术学院李伟等。

　　限于编写时间和编者的水平，书中难免存在不足之处，希望广大读者批评指正。

　　服务邮箱：476371891@qq.com。

编　者
2019 年 12 月

目 录

单元一	Android 系统介绍 ················ 1
1.1	Android 基本概述 ··················· 2
1.1.1	Android 简介 ···················· 2
1.1.2	Android 发展史 ·················· 3
1.1.3	Android 的系统架构 ············· 4
1.1.4	Android 环境 ··················· 6
1.2	SDK、AVD 简介 ····················· 10
1.3	编写 Android 应用 ················· 15
1.3.1	创建 Android 项目 ·············· 15
1.3.2	Android 模式的项目结构 ······ 16
1.3.3	运行 Android 项目 ·············· 18
1.3.4	app 目录下的结构 ··············· 19
【单元小结】 ····························· 20
【上机实战】 ····························· 20
【拓展作业】 ····························· 20

单元二 UI 布局详解 ···················· 21
  2.1 文本框(TextView) ···················· 22
    2.1.1 TextView基本用法 ············· 22
    2.1.2 TextView实现跑马灯效果 ···· 25
    2.1.3 代码设置TextView的属性 ··· 26
  2.2 编辑框(EditText) ···················· 27
    2.2.1 EditText 的用法 ················ 27
    2.2.2 获取 EditText 输入的值 ······ 31
  2.3 按钮(Button) ·························· 32
    2.3.1 Button 基本用法 ··············· 32
    2.3.2 Button 单击事件 ··············· 33

  2.4 提示框(Toast) ························· 37
    2.4.1 默认效果 ······················· 37
    2.4.2 自定义显示位置效果 ········· 38
    2.4.3 带图片效果 ···················· 39
  2.5 ImageView ····························· 40
  2.6 ImageButton ··························· 41
  2.7 日期和时间框(DatePicker、
        TimePicker) ·························· 42
  【单元小结】 ····························· 44
  【单元自测】 ····························· 45
  【上机实战】 ····························· 45
  【拓展作业】 ····························· 48

单元三 Activity 和 Intent ············· 49
  3.1 Activity 的介绍 ······················· 50
    3.1.1 Activity 的概念 ················ 50
    3.1.2 活动的基本用法 ··············· 50
    3.1.3 创建活动 ······················· 51
    3.1.4 创建布局 ······················· 53
    3.1.5 AndroidManifest 文件中
           注册 Activity ··················· 54
  3.2 Activity 的生命周期 ················· 56
  3.3 Intent 介绍和使用 ···················· 59
    3.3.1 显式 Intent ····················· 60
    3.3.2 隐式 Intent ····················· 61
  3.4 Intent 传值和传对象 ················· 62
    3.4.1 Intent 传值 ····················· 62

		3.4.2	Intent 传对象	63
		【单元小结】		65
		【单元自测】		65
		【上机实战】		65
		【拓展作业】		68
单元四	Android UI 布局详解			69
4.1	Android UI 布局介绍			70
4.2	Android UI 常用六大布局			71
		4.2.1	LinearLayout 线性布局详解	71
		4.2.2	RelativeLayout 相对布局详解	75
		4.2.3	FrameLayout 框架布局详解	80
		4.2.4	AbsoluteLayout 绝对布局详解	81
		4.2.5	TableLayout 表格布局详解	82
		4.2.6	GridLayout 网格布局详解	84
		【单元小结】		87
		【单元自测】		87
		【上机实战】		88
		【拓展作业】		88
单元五	Handler 消息传递机制			89
5.1	Handler 消息传递机制介绍			90
5.2	Handler 的使用			91
		【单元小结】		93
		【单元自测】		93
		【上机实战】		94
		【拓展作业】		94
单元六	UI 中级控件			95
6.1	RadioButton、RadioGroup			96
6.2	CheckBox			99
6.3	对话框(Dialog)			104
6.4	Spinner			108
6.5	ListView			111
		6.5.1	简介	111

		6.5.2	工作原理	111
		6.5.3	具体使用	113
		6.5.4	Adapter 介绍	115
		6.5.5	常用适配器介绍	116
6.6	GridView			129
6.7	ProgressBar			133
		【单元小结】		136
		【单元自测】		136
		【上机实战】		137
		【拓展作业】		142
单元七	Android 高级控件			143
7.1	Toolbar			144
		7.1.1	了解 Toolbar	144
		7.1.2	使用 Toolbar	145
7.2	ViewPager 和 PagerAdapter			149
		7.2.1	概述	149
		7.2.2	ViewPager 的重要属性	149
		7.2.3	PagerAdapter	150
		7.2.4	FragmentPagerAdapter 和 FragmentStatePagerAdapter	152
7.3	Fragment			153
		7.3.1	Fragment 概述	153
		7.3.2	Fragment 的生命周期图	154
		7.3.3	Fragment 的使用	154
		7.3.4	Fragment 管理与 Fragment 事务	156
		7.3.5	Fragment 与 Activity 的交互	157
7.4	TabLayout 导航栏			158
		7.4.1	TabLayout 的常用属性	159
		7.4.2	各种使用场景	159
		【单元小结】		163
		【单元自测】		163
		【上机实战】		163
		【拓展作业】		166
单元八	Service 的生命周期			167
8.1	Service 简介			168

8.2 Service 生命周期 ·············· 168
　　8.2.1 　创建 Service ·············· 171
　　8.2.2 　startService()启动服务 ···· 171
　　8.2.3 　bindService()绑定服务 ···· 174
8.3 简易音乐播放器 ·············· 178
　　【单元小结】 ·················· 180
　　【单元自测】 ·················· 180
　　【上机实战】 ·················· 181
　　【拓展作业】 ·················· 184

单元九　广播和通知 ·············· 185
9.1 广播机制简介 ················ 186
9.2 系统广播事件使用 ············ 187
9.3 自定义 BroadcastReceiver ····· 188
9.4 通知 ························ 190
　　【单元小结】 ·················· 194
　　【单元自测】 ·················· 194
　　【上机实战】 ·················· 195
　　【拓展作业】 ·················· 198

单元十　Android 中的数据存储 ···· 199
10.1 File 文件存储 ··············· 200
　　10.1.1 　Android 文件的操作
　　　　　　模式 ················ 200
　　10.1.2 　文件操作常用的 XML
　　　　　　属性 ················ 200
　　10.1.3 　文件读写的案例实现 ··· 201
10.2 SharedPreferences ·········· 205
　　10.2.1 　SharedPreferences简介 ··· 205
　　10.2.2 　SharedPreferences 的
　　　　　　使用模式 ············ 206
　　10.2.3 　SharedPreferences 的
　　　　　　使用案例 ············ 206
10.3 SQLite 数据库 ·············· 209
　　10.3.1 　SQLite 简介 ·········· 209
　　10.3.2 　SQLite 数据库的基础
　　　　　　知识 ················ 209
　　10.3.3 　SQLite 创建、打开
　　　　　　数据库及表 ·········· 210

　　10.3.4 　SQLite 案例实现 ······ 212
　　【单元小结】 ·················· 217
　　【单元自测】 ·················· 217
　　【上机实战】 ·················· 218
　　【拓展作业】 ·················· 224

单元十一　ContentProvider ······ 225
11.1 ContentProvider 概述 ······· 226
11.2 Uri 和 UriMatcher ·········· 227
　　11.2.1 　Uri介绍 ············· 227
　　11.2.2 　使用UriMatcher ······ 228
　　11.2.3 　ContentUris类及数据的
　　　　　　共享 ················ 229
11.3 操作 ContentProvider ······· 230
　　11.3.1 　使用 ContentResolver
　　　　　　操作 ContentProvider
　　　　　　中的数据 ············ 230
　　11.3.2 　通讯录操作 ·········· 231
11.4 自定义 ContentProvider ····· 233
　　【单元小结】 ·················· 245
　　【单元自测】 ·················· 245
　　【上机实战】 ·················· 245
　　【拓展作业】 ·················· 246

单元十二　网络编程 ·············· 247
12.1 网络编程概述 ··············· 248
　　12.1.1 　使用 HTTP ··········· 248
　　12.1.2 　Android 网络程序的
　　　　　　功能 ················ 248
12.2 HTTP 网络编程 ············· 249
12.3 URL 网络编程 ·············· 251
12.4 WebView ··················· 253
　　12.4.1 　WebView 介绍 ········ 253
　　12.4.2 　WebView 的作用 ······ 253
　　12.4.3 　WebView 的使用 ······ 254
　　12.4.4 　WebView 与 H5 混合
　　　　　　开发 ················ 255
12.5 网络框架 ··················· 266
　　12.5.1 　网络框架简介 ········ 266

12.5.2 okhttp 框架的优点和
作用·····················267
12.5.3 okhttp 的使用·············267
【单元小结】·····················278
【单元自测】·····················278
【上机实战】·····················278
【拓展作业】·····················282

单元十三 XML/JSON 数据解析·········283

13.1 XML 数据解析·················284
13.1.1 Pull 解析方式···········284
13.1.2 DOM 解析方式··········287
13.2 JSON 数据解析················289
13.2.1 基础结构················290
13.2.2 JSON 数据解析·········290
【单元小结】·····················298
【单元自测】·····················298
【上机实战】·····················298
【拓展作业】·····················301

单元一 Android 系统介绍

课程目标

- Android 基本概述
- Android 配置
- Android 应用程序

 简介

在Google及其开放手机联盟推出基于Linux平台的开源手机操作系统Android之后，Google又不惜重金举办了Android开发者大赛，吸引了众多开发者的目光。Android不仅功能强大，而且具有开放和免费等先天优势，全球范围内的电信行业、手机制造商因此毫不犹豫地加入Android开放手机联盟中来。

随着Android手机的普及，Android应用的需求势必越来越大，这将是一个潜力巨大的市场，会吸引无数开发厂商和开发者投身其中。作为程序员的我们，当然不应该落后于人，赶快加入Android的阵营中来吧！

1.1 Android基本概述

Android是一种基于Linux的自由及开放源代码的操作系统，主要应用于移动设备，如智能手机和平板电脑，由Google(谷歌)公司和开放手机联盟领导及开发，尚未有统一中文名称，中国大陆地区较多人称其为"安卓"或"安致"。Android操作系统最初由Andy Rubin开发，主要支持手机。2005年8月由Google收购注资。2007年11月，Google与84家硬件制造商、软件开发商及电信营运商组建开放手机联盟共同研发改良Android系统。随后Google以Apache开源许可证的授权方式，发布了Android的源代码。第一部Android智能手机发布于2008年10月。Android逐渐扩展到平板电脑及其他领域上，如电视机、数码相机、游戏机、智能手表等。2011年第一季度，Android在全球的市场份额首次超过Symbian(塞班)系统，跃居全球第一。2013年的第四季度，Android平台手机的全球市场份额已经达到78.1%。2013年9月24日，在Android迎来5岁生日之际，全世界采用这款系统的设备数量已经达到10亿台。2015年年末，Android占据全球智能手机操作系统市场80%的份额，占据中国智能手机操作系统市场90%的份额。

1.1.1 Android简介

Android一词的本义指"机器人"，同时也是Google于2007年11月5日宣布的基于Linux平台的开源手机操作系统的名称，该平台由操作系统、中间件、用户界面和应用软件组成。

Android系统具有如下5个特点。

- 开放性：Google与开放手机联盟合作开发了Android，Google通过运营商、设备制造商、开发商和其他有关方结成深层次的合作伙伴，希望通过建立标准化、开放式的移动电话平台，在移动产业内形成一个开放式的生态系统。
- 应用程序列界限：Android上的应用程序可以通过标准API访问核心移动设备功能。通过互联网，应用程序可以声明它们的功能可供其他程序调用。

- 应用程序是在平等条件下创建的：移动设备上的应用程序可以替换或扩展，即使是拨号程序或主屏幕这样的核心组件。
- 应用程序可以轻松地嵌入网络：应用程序可以轻松地嵌入 HTML、JavaScript 和样式表，还可以通过 WebView 显示网络内容。
- 应用程序可以并行运行：Android 是一个完整的多任务环境，应用程序可以在其中并行运行。在后台运行时，应用程序可以生成通知引起注意。

1.1.2 Android 发展史

2007 年 11 月 5 日，开放手机联盟成立。

2007 年 11 月 12 日，Google 发布 Android SDK 预览版，这是第一个对外公布的 Android SDK，为发布正式版收集用户反馈。

2008 年 4 月 17 日，Google 举办开发者竞赛。

2008 年 8 月 28 日，Google 开通 Android Market，供 Android 手机下载需要使用的应用程序。

2008 年 9 月 23 日，发布 Android SDK v1.0 版，这是第一个稳定的 SDK 版本。

2008 年 10 月 21 日，Google 开放 Android 平台的源代码。

2008 年 10 月 22 日，第一款 Android 手机 T-Mobile G1 在美国上市，由中国台湾宏达国际电子公司(简称宏达电)制造。

2009 年 2 月，发布 Android SDKv1.1 版。

2009 年 2 月 17 日，第二款 Android 手机 T-Mobile G2 正式发售，仍由中国台湾宏达电制造。

2009 年 4 月 15 日，发布 Android SDK v1.5 版。

2009 年 6 月 24 日，中国台湾宏达电发布了第三款 Android 手机 HTC Hero，如图 1-1 所示。

2010 年 5 月，Google 正式发布了 Android 2.2 操作系统。

2010 年 12 月，Google 正式发布了 Android 2.3 操作系统 Gingerbread(姜饼)。

2011 年 9 月，Google 发布全新的 Android 4.0 操作系统，这款系统被 Google 命名为 Ice Cream Sandwich(冰激凌三明治)。

2013 年 11 月 1 日，Android 4.4 正式发布，从具体功能上讲，Android 4.4 提供了各种实用小功能，新的 Android 系统更智能，添加了更多的 Emoji 表情图案，UI 的改进也更现代，如全新的 Hello iOS7 半透明效果。

2014 年 10 月 15 日(美国太平洋时间)，发布了全新 Android 操作系统——Android 5.0。

2015 年 5 月 28 日，Google I/O 2015 大会举行。在发布会上代号为 Marshmallow(棉花糖)的 Android 6.0 系统正式推出。

2016 年 5 月 18 日，Google 2016 年的 I/O 开发者大会举行，Android 7.0 系统发布(代号为 Android N)。

图 1-1

Android 提供访问硬件的 API 函数,简化了摄像头、GPS 等硬件的访问过程;具有自己的运行时和虚拟机;提供丰富的界面控件供使用者之间调用,加快用户界面的开发速度,保证 Android 平台上程序界面的一致性;提供轻量级的进程间通信机制 Intent,使跨进程组件通信和发送系统级广播成为可能;提供了 Service 作为无用户界面、长时间后台运行的组件;支持高效、快速的数据存储方式。

表 1-1 列出了目前主要的 Android 系统版本及其详细信息,当我们看到这张表格时,数据可能已经发生了变化。

表 1-1

版本号	系统代码	API	市场占有率/%
2.2	Froyo	8	0.1
2.3.3～2.3.7	Gingerbread	10	1.5
4.0.3～4.0.4	IceCreamSandwich	15	1.3
4.1～4.3	Jelly Bean	16,17,18	15.6
4.4	KitKat	19	27.7
5.0～5.1	Lollipop	21,22	35
6.0	Marshmallow	23	18.7

1.1.3 Android 的系统架构

Android 采用软件分层的架构,共分为四层,如图 1-2 所示。

1. Linux 内核(见图 1-3)

(1) 硬件和其他软件层之间的一个抽象隔离层。
(2) 提供安全机制、内存管理、进程管理、网络协议栈和驱动程序等。

2. 中间件层(见图 1-4)

(1) 由函数库和 Android 运行时构成。
(2) 函数库,主要提供一组基于 C/C++的函数库。
(3) Surface Manager,支持显示子系统访问,提供应用程序与 2D、3D 图像层平滑连接。

(4) Media Framework，实现音视频的播放和录制功能。
(5) SQLite，轻量级的关系数据库引擎。
(6) OpenGL | ES，基于 3D 图像加速。
(7) FreeType，位图与矢量字体渲染。
(8) WebKit，Web 浏览器引擎。
(9) SGL，2D 图像引擎。
(10) SSL，数据加密与安全传输的函数库。
(11) Libc，标准 C 运行库，Linux 系统中底层应用程序开发接口。
(12) 核心库，提供 Android 系统的特有函数功能和 Java 语言函数功能。
(13) Dalvik 虚拟机，实现基于 Linux 内核的线程管理和底层内存管理。

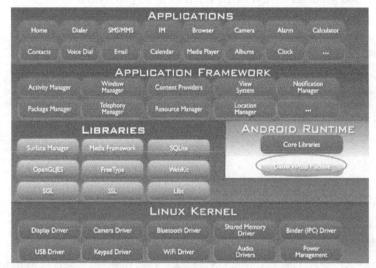

图 1-2

图 1-3

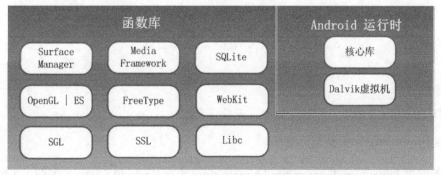

图 1-4

3. 应用程序框架(见图 1-5)

(1) 提供 Android 平台基本的管理功能和组件重用机制。
(2) Activity Manager，管理应用程序的生命周期。
(3) Windows Manager，启动应用程序的窗体。
(4) Content Provider，共享私有数据，实现跨进程的数据访问。
(5) Package Manager，管理安装在 Android 系统内的应用程序。
(6) Telephony Manager，管理与拨打、接听电话相关的功能。
(7) Resource Manager，允许应用程序使用非代码资源。
(8) Location Manager，管理地图相关的服务功能。
(9) Notification Manager，允许应用程序在状态栏中显示信息。
(10) ViewSystem，构建应用程序的基本组就是文本框、按钮等。

图 1-5

1.1.4 Android 环境

配置 Android 开发环境之前，首先需要了解 Android 对操作系统的要求。Android 可以应用于 Windows XP 及以上版本、MacOS、Linux 等操作系统，本书以 Windows 10 为操作系统，以 Android Studio 为开发工具进行讲解。Android 开发所需软件版本及下载地址如下。

1. Java JDK 下载

进入网页 http://java.sun.com/javase/downloads/index.jsp，如图 1-6 所示。

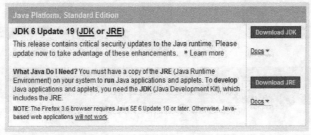

图 1-6

选择 Download JDK，只下载 JDK，无须下载 JRE。

2. Android Studio 下载与安装

进入网页 https://developer.android.google.cn/，选择 Android Studio，如图 1-7 所示。

图 1-7

我们下载下来的将是一个安装包，安装的过程也很简单，逐次单击 Next 按钮即可。其中，选择安装组件时建议全部勾选，如图 1-8 所示。

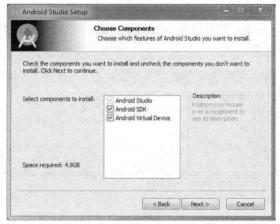

图 1-8

接下来，选择 Android Studio 及 Android SDK 的安装地址，这些可根据自己计算机的实际情况进行选择，若不想改动，则保持默认设置即可，如图 1-9 所示。

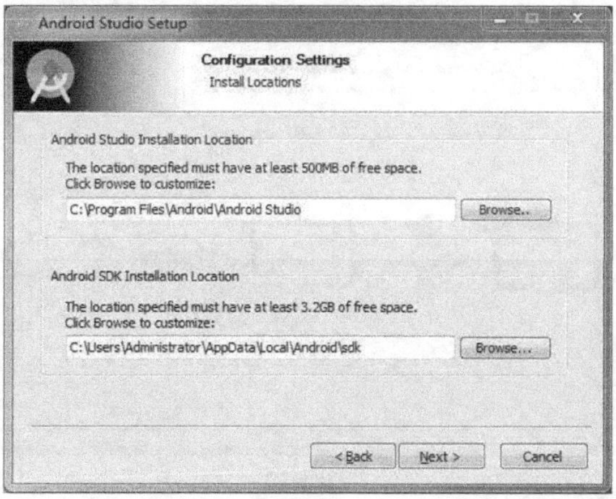

图 1-9

选择安装地址后，其他项全部保持默认设置，逐次单击 Next 按钮即可完成安装，如图 1-10 所示。

图 1-10

安装完成后，单击 Finish 按钮启动 Android Studio，一开始会让我们选择是否导入之前 Android Studio 版本的配置，因为是首次安装，所以这里选择不导入，如图 1-11 所示。

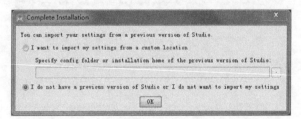

图 1-11

选择不导入配置后，单击 OK 按钮，进入 Android Studio 的配置界面，如图 1-12 所示。

图 1-12

进入 Android Studio 的配置界面后，单击 Next 按钮，进行具体的配置，如图 1-13 所示。

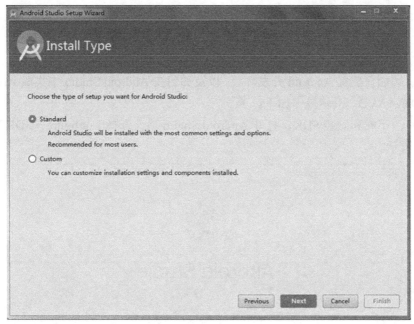

图 1-13

安装类型可以选择 Android Studio，有 Standard 和 Custom 两种。Standard 表示一切都使用默认的配置，比较方便；Custom 则可以根据用户的特殊需求进行自定义。简单起见，这里我们选择 Standard 类型，继续单击 Next 按钮，完成配置工作。

单击 Finish 按钮，配置工作全部完成，Android Studio 会尝试联网下载一些更新，更新完成后，单击 Finish 按钮，进入 Android Studio 的欢迎界面，如图 1-14 所示。

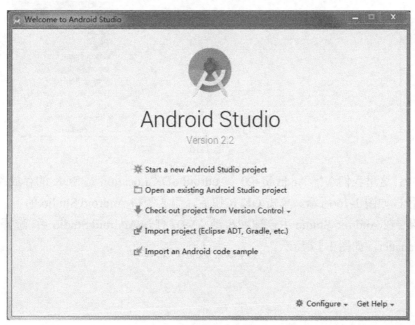

图 1-14

至此，Android Studio 安装完成。同理，下载并安装 Android SDK。

1.2 SDK、AVD 简介

下面讲解如何配置 Android 开发环境，以及如何在 Android Studio 中集成 Android 开发所需的 SDK、AVD，具体操作如下。

第一步：下载和安装 SDK，打开 Android Studio 后，选择 Configure→SDK Manager，如图 1-15 所示。

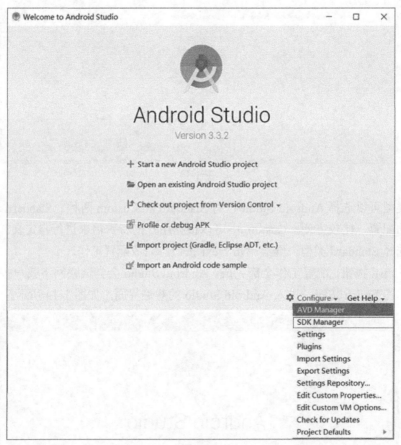

图 1-15

第二步：这里我们选择 Android 6.0，Android SDK Location 是 SDK 的存放路径，可自由选择路径，如图 1-16 所示。单击 OK 按钮后，耐心等待 Android Studio 的下载和安装。

第三步：在 Android Studio 中创建和启动 AVD，打开 Android Studio 后，选择 Configure→AVD Manager，如图 1-17 所示。

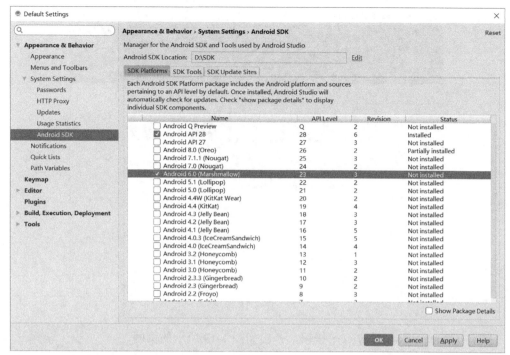

图 1-16

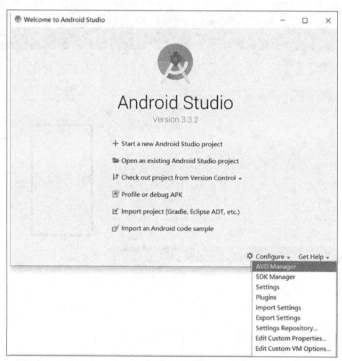

图 1-17

第四步：单击 AVD Manager，进行操作，如图 1-18 所示。

图 1-18

单击+Create Virtual Device…按钮，弹出如图 1-19 所示的页面。

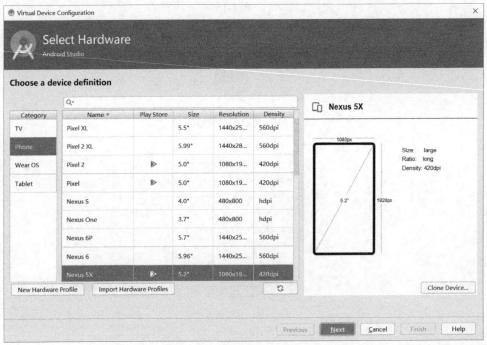

图 1-19

选择页面左边的 Phone，右边的机型可以任意选择，单击 Next 按钮，弹出选择 System Image(系统镜像)窗口，可以随意选择(若选择未下载，则 Next 按钮不可用)，然后单击 Download 按钮开始安装，如图 1-20 和图 1-21 所示。

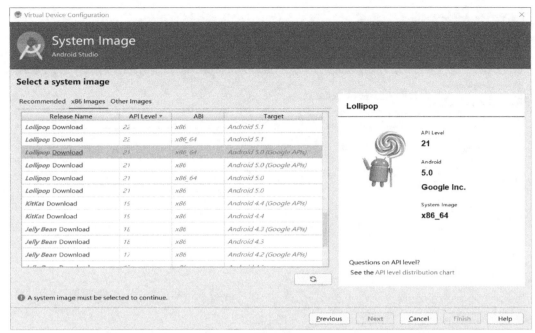

图 1-20

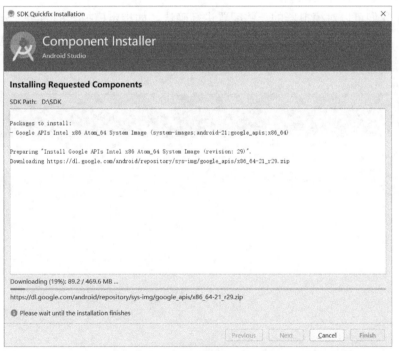

图 1-21

第五步：下载完成后，回到图 1-20 所示的界面，选中下载完毕的 System Image 即可单击 Next 按钮，打开如图 1-22 所示的界面。

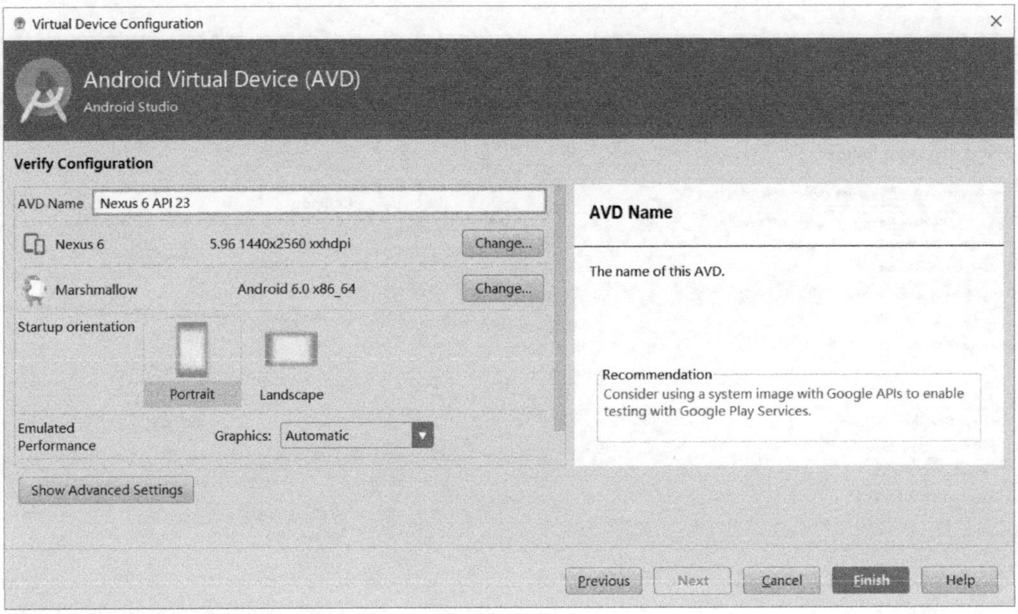

图 1-22

第六步：确认模拟器配置。

在这里我们可以对模拟器的一些配置进行确认，如指定模拟器的名字、分辨率、横竖屏等信息，如果没有特殊需求，全部保持默认设置即可。单击 Finish 按钮，完成模拟器的创建，弹出如图 1-23 所示的窗口。

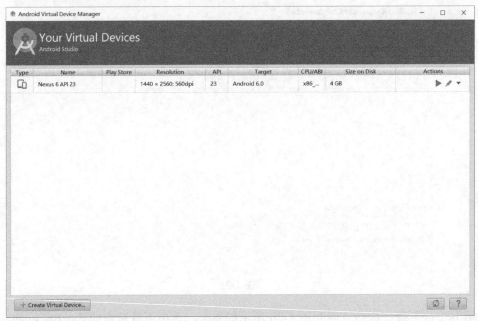

图 1-23

可以看到，现在模拟器列表中已经存在一个创建好的模拟器设备了，单击 Actions 栏目中最左边的三角形按钮即可启动模拟器。模拟器会像手机一样，有一个开机过程，启动完成之后的界面如图 1-24 所示。

图 1-24

至此，Android 开发环境就已经全部搭建完成了。现在应该做什么？当然是写下你的第一行 Android 代码了，让我们快点开始吧。

1.3　编写 Android 应用

任何一个编程语言写出的第一个程序毫无疑问都会是 Hello World，这是自 20 世纪 70 年代一直流传下来的传统，在编程界已成为永恒的经典。

1.3.1　创建 Android 项目

在 Android Studio 的欢迎界面中，单击 Start a new Android Studio project，打开一个创建新项目的界面，如图 1-25 所示。

其中，Application name 表示应用名称，此应用安装到手机上后会在手机上显示该名称，这里我们填入 HelloWorld。

Package name 表示项目的包名，Android 系统就是通过包名来区分不同应用程序的，因此包名一定要具有唯一性。Android Studio 会根据应用名称自动帮我们生成合适的包名，如果我们不想使用默认生成的包名，可以单击右侧的 Edit 按钮自行修改。

Project location 表示项目代码存放的位置，如果没有特殊要求，保持默认即可。

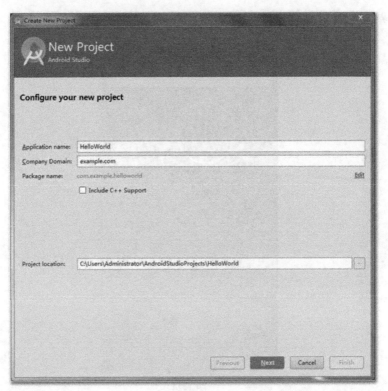

图 1-25

1.3.2　Android 模式的项目结构

在创建好项目之后，会进入如图 1-26 所示的页面。

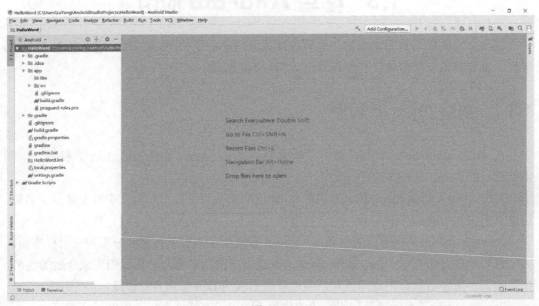

图 1-26

任何一个新建的项目都会默认使用 Android 模式的项目结构，但这并不是项目真实的目录结构，而是被 Android Studio 转换过的。这种项目结构简洁明了，适合进行快速开发，但是对于新手来说可能并不易于理解。单击图 1-26 中的 Android 区域可以切换项目结构模式，如图 1-27 所示。

这里我们将项目结构模式切换成 Project，这就是项目真实的目录结构了，如图 1-28 所示。

图 1-27

图 1-28

Project 模式的项目结构

一开始看到这么多陌生的东西，我们一定会感到头晕目眩。别担心，现在就对图 1-28 中的内容进行一一讲解，之后再看这张图就不会感到那么吃力了。

- .gradle 和 .idea 两个目录下放置的都是 Android Studio 自动生成的一些文件，我们无须关心，也不要去手动编辑。
- app：项目中的代码、资源等内容几乎都是放置在该目录下，我们后面的开发工作也基本都是在这个目录下进行的，下面还会对该目录展开讲解。
- gradle：这个目录下包含 gradle wrapper 的配置文件，使用 gradle wrapper 的方式不需要提前下载 gradle，系统会自动根据本地的缓存情况来决定是否需要联网下载。Android Studio 默认没有启用 gradle wrapper 的方式，如果需要打开，可以单击 Android Studio 导航栏→File→Settings→Build→Execution→Deployment→Gradle 进行配置更改。
- .gitignore：该文件是用来将指定的目录或文件排除在版本控制之外的。
- build.gradle：这是项目全局的 gradle 构建脚本，通常这个文件中的内容是不需要修改的。下面我们将会详细分析 gradle 构建脚本中的具体内容。
  - gradle.properties：该文件是全局的 gradle 配置文件，在这里配置的属性将会影响项目中所有的 gradle 编译脚本。
  - gradlew 和 gradlew.bat：这两个文件是用来在命令行界面中执行 gradle 命令的，其中，gradlew 是在 Linux 或 Mac 系统中使用的，gradlew.bat 是在 Windows 系统中使用的。

- HelloWorld.iml：该文件是所有 IntelliJ IDEA 项目都会自动生成的一个文件（Android Studio 是基于 IntelliJ IDEA 开发的），用于标识这是一个 IntelliJ IDEA 项目，不需要修改这个文件中的任何内容。
- local.properties：该文件用于指定本机中的 Android SDK 路径，通常内容都是自动生成的，不需要修改，如果我们本机中的 Android SDK 位置发生了变化，那么可将这个文件中的路径改成新的位置。
- settings.gradle：该文件用于指定项目中所有引入的模块。由于 Hello World 项目中就只有一个 app 模块，所以该文件中也就只引入了 app 这一个模块。通常情况下，模块的引入都是自动完成的，很少需要我们手动去修改这个文件的场景。

现在整个项目的外层目录结构已经介绍完了。我们会发现，除了 app 目录之外，大多数的文件和目录都是自动生成的，不需要进行修改，因此，app 目录下的内容才是我们以后的工作重点，展开之后的结构如图 1-29 所示。

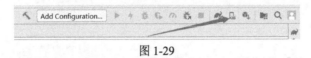

图 1-29

1.3.3 运行 Android 项目

在运行项目之前，要先打开虚拟机，如果我们不小心关掉了前面运行的虚拟机，那么可以从图 1-30 所示的页面中打开。

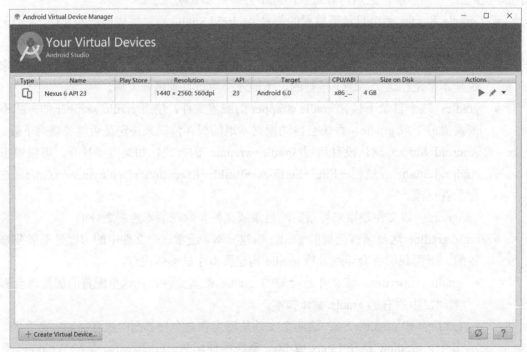

图 1-30

1.3.4　app 目录下的结构

下面我们对 app 目录下的内容(见图 1-31)进行更为详细的分析。

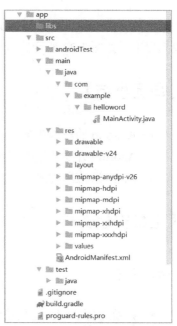

图 1-31

- libs：如果我们的项目中使用到了第三方 jar 包，则需要把这些 jar 包都放在 libs 目录下，而且放在这个目录下的 jar 包都会被自动添加到构建路径中。
- androidTest：此处是用来编写 AndroidTest 测试用例的，可以对项目进行一些自动化测试。
- java：毫无疑问，java 目录是放置我们所有 Java 代码的地方，展开该目录，我们将在里面看到刚才创建的 Hello World Activity 文件。
- res：该目录下的内容较多。简单来说，就是我们在项目中使用到的所有图片、布局、字符串等资源都要存放在这个目录下。该目录下还有很多子目录，如图片放在 drawable 目录下，布局放在 layout 目录下，字符串放在 values 目录下，所以我们不用担心会让整个 res 目录杂乱无章。
- AndroidManifest.xml：这是整个 Android 项目的配置文件，我们在程序中定义的所有四大组件都需要在这个文件里注册，另外，还可以在这个文件中给应用程序添加权限声明。由于这个文件以后会经常用到，届时再做详细说明。
- test：此处用来编写 UnitTest 测试用例，是对项目进行自动化测试的另一种方式。
- .gitignore：该文件用于将 app 模块内的指定目录或文件排除在版本控制之外，作用与外层的.gitignore 文件类似。
- build.gradle：这是 app 模块的 gradle 构建脚本，该文件中会指定很多项目构建相关的配置。
- proguard-rules.pro：该文件用于指定项目代码的混淆规则，当代码开发完成后打成安装包文件，如果不希望代码被别人破解，通常会将代码进行混淆，从而让破解者难以阅读。

这样整个项目的目录结构就介绍完了，如果我们还不能完全理解也很正常，毕竟里面有太多的内容都还有没接触过，但这并不会影响我们后面的学习。等我们学完整本书再回来看这个目录结构图时，就会觉得特别的清晰和简单。

运行 Hello World 项目

通常，app 就是当前的主项目，右边的三角形按钮是用来运行项目的，如图 1-32 所示。

图 1-32

稍微等待一会儿，Hello World 项目就会运行到模拟器上，如图 1-33 所示。

【单元小结】

- Android 的基本概念。
- Android 开发环境的配置。
- Android 应用程序。

【上机实战】

上机目标

- 熟练掌握 Android 开发环境的搭建步骤。
- 使用 Android 进行简单页面的开发。

图 1-33

上机练习

练习：编写一个"为自己代言"的程序

【问题描述】
今天我们就不写 Hello World 了，写一个"为自己代言"。

【问题分析】
首先搭建开发环境，然后启动 Android 模拟器。

【参考步骤】
创建 Android 工程。请参考本单元前面讲述的步骤。

【拓展作业】

1. 熟悉如何在 MyEclipse 中集成 SDK、AVD。
2. 请按照开发环境搭建的步骤编写"我爱我的爸爸妈妈"程序。

单元二 UI 布局详解

 课程目标

- ▶ Android UI 简单布局
- ▶ Activity UI 常用控件
- ▶ Activity UI 常用控件的使用

 **简 介**

布局(layout)的概念是针对 Activity 的，Activity 就是布满整个 Android 设备的窗口或者悬浮于其他窗口上的交互界面。通常情况下，开发人员可以使用两种方式来创建 UI 组件，一种是使用 XML 方式来配置 UI 组件的相关属性，然后装载这些 UI 组件，这也是最常用的方式。但是有些特殊情况下，需要动态生成 UI 组件，这就需要使用第二种方式，即完全使用 Java 代码来创建 UI 组件。本单元将重点介绍 UI 布局。

应用程序的人机交互界面由很多 Android 控件组成，前面章节所有示例的界面都是由一些常用的控件构成。Android 的控件可以说是目前所有手机平台控件中最为完美的，今天开始将对这些控件进行详细介绍。本单元讲解如图 2-1 所示的控件。

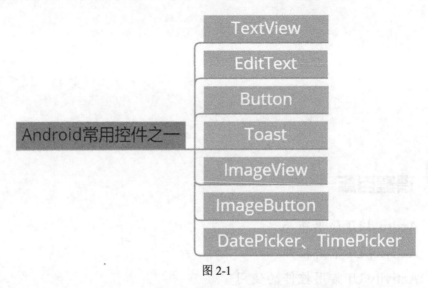

图 2-1

2.1 文本框(TextView)

在界面上需要显示文本内容，就用到了 TextView，而 TextView 在 Android 中可以理解为一个文本视图控件，通常会在.XML 布局文件中为文本视图控件指定各种属性来设置它的样式。下面阐述 TextView 的基本用法。

2.1.1 TextView 基本用法

我们可以用 TextView 显示一段文本内容，如图 2-2 所示。

图 2-2

下面看一看它的源代码：

```
<LinearLayout xmlns:android="http://schemas.android.com/apk/res/android"
    xmlns:tools="http://schemas.android.com/tools"
    android:layout_width="match_parent"
    android:layout_height="match_parent"
    android:orientation="vertical"
    android:padding="10dip"
    tools:context="com.hp.main.MainActivity" >

    <TextView
        android:layout_width="match_parent"
        android:layout_height="wrap_content"
        android:text="你好！Android"/>

</LinearLayout>
```

以上只是 TextView 常用的、简单的属性，如果我们要设置 TextView 字体的各种样式，应该如何做呢？表 2-1 所示是 TextView 的一些基本属性及说明。

表 2-1

属性	说明
android:width	设置文本区域的宽度，只支持度量单位，如 px(像素)、dp、sp、in、mm(毫米)
android:layout_width	设置文本区域的宽度，一般是"fill_parent","wrap_content","match_parent"。当然，它也可以像前者一样，设置数值
android:maxLength	限制输入字符数
android:lines	设置文本的行数，设置两行就显示两行，即使第二行没有数据
android:lineSpacingExtra	设置行间距

(续表)

属性	说明
android:textAppearance	设置文字外观。如?android:attr/textAppearanceLargeInverse 中引用的是系统自带的一个外观，?表示系统是否有这种外观，否则使用默认的外观
android:textColor	设置文本颜色
android:textScaleX	设置文字之间间隔，默认为 1.0f
android:textStyle	设置字形[bold(粗体)、italic(斜体)、bolditalic(又粗又斜)]
android:password	以小点"."显示文本

在实际操作中，表 2-1 中的属性基本能够满足用户的需要，如使用 android:textColor 在 XML 中就能够设置想要的颜色，使用 android:textSize 在 XML 中就能够设置字体的大小，如图 2-3 所示。

图 2-3

下面看一看它的源代码：

```
<LinearLayout xmlns:android="http://schemas.android.com/apk/res/android"
    xmlns:tools="http://schemas.android.com/tools"
    android:layout_width="match_parent"
    android:layout_height="match_parent"
    android:orientation="vertical"
    android:padding="10dip"
    tools:context="com.hp.main.MainActivity" >

    <TextView
        android:layout_width="match_parent"
        android:layout_height="wrap_content"
        android:text="你好！Android"
        android:textSize="20sp"
        android:textColor="@color/textColor"/>

</LinearLayout>
```

在 Android 中字体大小使用的单位是 sp，在这里设置字体大小 textSize 的属性值为20sp。同时，这里也设置了 textColor 的属性，只是颜色的值放在了 color.xml 的文件中。

2.1.2 TextView 实现跑马灯效果

我们经常会看到网上有一些跑马灯的效果，其实 TextView 就可以直接实现该功能，只需要设置它的配置就可以了，如图2-4 所示。

图 2-4

源代码如下：

```
<LinearLayout xmlns:android="http://schemas.android.com/apk/res/android"
    xmlns:tools="http://schemas.android.com/tools"
    android:layout_width="match_parent"
    android:layout_height="match_parent"
    android:orientation="vertical"
    android:padding="10dip"
    tools:context="com.hp.ui.TextViewActivity" >

    <TextView
        android:layout_width="match_parent"
        android:layout_height="wrap_content"
        android:textStyle="normal"
        android:textSize="16dp"
        android:text="@string/marqueeText"
        android:ellipsize="marquee"
        android:focusable="true"
        android:focusableInTouchMode="true"
        android:singleLine="true"
        android:layout_margin="10dp"/>

</LinearLayout>
```

以上代码中，我们首先要通过 singleLine 属性设置文本只能显示一行，然后将 ellipsize 的属性设置为 marquee，因为 TextView 默认是不获取焦点的，所以还需要将 focusable 设置为 true。

2.1.3 代码设置 TextView 的属性

有些时候为了更好地、灵活地设置 TextView 的属性，可以直接在代码中操作，而不需要在 XML 中做任何设置。代码如下。

activity_main.xml：

```xml
<LinearLayout xmlns:android="http://schemas.android.com/apk/res/android"
    xmlns:tools="http://schemas.android.com/tools"
    android:layout_width="match_parent"
    android:layout_height="match_parent"
    android:orientation="vertical"
    android:padding="10dip"
    tools:context="com.hp.mian.MainActivity" >

    <TextView
        android:id="@+id/textView"
        android:layout_width="match_parent"
        android:layout_height="wrap_content"
        android:text=""
        android:textSize="20sp"
        android:textColor="@color/textColor"/>

</LinearLayout>
```

MainActivity.java：

```java
import android.graphics.Color;
import android.graphics.Color;
import android.os.Bundle;
import android.support.annotation.Nullable;
import android.support.v7.app.AppCompatActivity;
import android.widget.TextView;

import com.hp.main.R;

public class MainActivity extends AppCompatActivity {
    private TextView textView;
    @Override
    protected void onCreate(@Nullable Bundle savedInstanceState) {
        super.onCreate(savedInstanceState);
        setContentView(R.layout.activity_text4);
        textView = (TextView)findViewById(R.id.textView);
        //设置显示文本
        textView.setText("代码设置 TextView 的设置");
        //设置字体大小
        textView.setTextSize(20);
        //设置文字的颜色
```

```
            textView.setTextColor(Color.RED);
    }
}
```

结果如图 2-5 所示。

图 2-5

2.2 编辑框(EditText)

EditText在开发中也是经常使用到的控件,例如,我们经常见到的登录界面的输入框,需要用到文本编辑输入框、密码编辑输入框,然后获得用户输入的内容并交给后台去处理。

2.2.1 EditText 的用法

EditText 的基本用法,如图 2-6 所示。

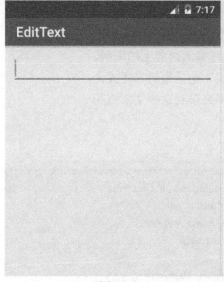

图 2-6

它是如何实现的呢？其代码如下。

activity_main.xml：

```
<LinearLayout xmlns:android="http://schemas.android.com/apk/res/android"
    xmlns:tools="http://schemas.android.com/tools"
    android:layout_width="match_parent"
    android:layout_height="match_parent"
    android:orientation="vertical"
    android:padding="10dip"
    tools:context="com.hp.main.MainActivity" >
    <EditText
        android:layout_width="match_parent"
        android:layout_height="wrap_content" />
</LinearLayout>
```

可以看到这里只使用了它的宽高属性，并没有使用其他属性，那么它还有哪些常用的属性呢？具体如下。

提示信息属性：

android:hint="我是提示信息"

限制 EditText 中输入的字符数量：

android:maxLength="11"

在 EditText 中设置输入的密码类型：

android:password="true"

在 EditText 中设置输入的数据类型：

android:inputType="number"

在 EditText 中设置输入的电话类型：

android:phoneNumber="true"

可以看到 EditText 也像 JSP 页面上的文本框一样，在界面上对文本框进行输入内容上的限制，如只能输入整数、邮箱格式、电话号码等样式。那么在 Android 中具体怎样来限制这些文本框的输入呢？如图 2-7 所示。

图 2-7

activity_main.xml：

```
<LinearLayout xmlns:android="http://schemas.android.com/apk/res/android"
    xmlns:tools="http://schemas.android.com/tools"
    android:layout_width="match_parent"
    android:layout_height="match_parent"
    android:orientation="vertical"
    android:padding="10dip"
    tools:context="com.hp.main.MainActivity" >
```

```xml
<LinearLayout
    android:layout_width="fill_parent"
    android:layout_height="wrap_content" >
    <TextView
        android:layout_width="wrap_content"
        android:layout_height="wrap_content"
        android:text="用户名" />
    <!-- 设置显示在控件上的提示信息(android:hint) -->
    <EditText
        android:id="@+id/txt1"
        android:layout_width="200dip"
        android:layout_height="wrap_content"
        android:hint="请输入用户名"
        android:textSize="14px" />
</LinearLayout>
<LinearLayout
    android:layout_width="fill_parent"
    android:layout_height="wrap_content" >
    <TextView
        android:layout_width="wrap_content"
        android:layout_height="wrap_content"
        android:text="密码" />
    <!-- 密码框( android:password="true") -->
    <EditText
        android:id="@+id/txt2"
        android:layout_width="200dip"
        android:layout_height="wrap_content"
        android:hint="请输入密码"
        android:password="true"
        android:textSize="14px" />
</LinearLayout>
<LinearLayout
    android:layout_width="fill_parent"
    android:layout_height="wrap_content" >
    <TextView
        android:layout_width="wrap_content"
        android:layout_height="wrap_content"
        android:text="年龄" />
    <!--设置只能输入整数(android:numeric=integer) -->
    <EditText
        android:id="@+id/txt3"
        android:layout_width="200dip"
        android:layout_height="wrap_content"
        android:hint="请输入年龄"
        android:numeric="integer"
        android:textSize="14px" />
</LinearLayout>
```

```xml
<LinearLayout
    android:layout_width="fill_parent"
    android:layout_height="wrap_content" >
    <TextView
        android:layout_width="wrap_content"
        android:layout_height="wrap_content"
        android:text="电话" />
    <!--输入电话号码(android:phoneNumber="true")   -->
    <EditText
        android:id="@+id/txt4"
        android:layout_width="200dip"
        android:layout_height="wrap_content"
        android:hint="请输入"
        android:phoneNumber="true"
        android:textSize="14px" />
</LinearLayout>
<LinearLayout
    android:layout_width="fill_parent"
    android:layout_height="wrap_content" >
    <TextView
        android:layout_width="wrap_content"
        android:layout_height="wrap_content"
        android:text="身份证" />
    <!--文本框是否可编辑(android:editable="false") -->
    <EditText
        android:id="@+id/txt5"
        android:layout_width="200dip"
        android:layout_height="wrap_content"
        android:hint="请输入身份证"
        android:editable="false"
        android:textSize="14px" />
</LinearLayout>
</LinearLayout>
```

EditText 编辑框的其他属性及说明如表 2-2 所示。

表 2-2

属性	说明
android:singleLine="true"	设置单行输入，一旦设置为 true，则文字不会自动换行
android:autoText	自动拼写帮助
android:capitalize	首字母大写
android:singleLine	是否单行或者多行，回车是离开文本框还是文本框增加新行
android:textColor	字体颜色
android:textSize	字体大小
android:textColorHint	设置提示信息文字的颜色，默认为灰色

2.2.2 获取 EditText 输入的值

现在我们可以使用 EditText 编辑框了，但在实际使用中往往需要获取用户输入 EditText 的值，那么怎么获取它的值呢？

activity_main.xm：

```xml
<LinearLayout xmlns:android="http://schemas.android.com/apk/res/android"
    xmlns:tools="http://schemas.android.com/tools"
    android:layout_width="match_parent"
    android:layout_height="match_parent"
    android:orientation="vertical"
    android:padding="10dip"
    tools:context="com.hp.main.MainActivity" >
    <LinearLayout
        android:layout_width="fill_parent"
        android:layout_height="wrap_content" >
      <TextView
          android:layout_width="wrap_content"
          android:layout_height="wrap_content"
          android:text="用户名" />
      <EditText
          android:id="@+id/username"
          android:layout_width="200dip"
          android:layout_height="wrap_content"
          android:hint="请输入用户名"
          android:textSize="16px" />
    </LinearLayout>
    <LinearLayout
        android:layout_width="fill_parent"
        android:layout_height="wrap_content" >
      <TextView
          android:layout_width="wrap_content"
          android:layout_height="wrap_content"
          android:text="密码" />
      <EditText
          android:id="@+id/password"
          android:layout_width="200dip"
          android:layout_height="wrap_content"
          android:hint="请输入密码"
          android:password="true"
          android:textSize="16px" />
    </LinearLayout>
</LinearLayout>
```

MainActivity.java：

```java
import android.os.Bundle;
```

```java
import android.support.annotation.Nullable;
import android.support.v7.app.AppCompatActivity;
import android.util.Log;
import android.widget.EditText;

import com.hp.main.R;

public class MainActivity extends AppCompatActivity{
    private EditText username;
    private EditText password;
    @Override
    protected void onCreate(@Nullable Bundle savedInstanceState) {
        super.onCreate(savedInstanceState);
        setContentView(R.layout.activity_main);
        username = (EditText)findViewById(R.id.username);
        password = (EditText)findViewById(R.id.password);
        String nameStr = username.getText().toString();
        String pwdStr = password.getText().toString();
        Log.d("MainActivity","用户名："+nameStr);
        Log.d("MainActivity","密码："+pwdStr);
    }
}
```

效果如图 2-8 所示。

```
D/MainActivity: 用户名:
D/MainActivity: 密码: 123
```

图 2-8

2.3 按钮(Button)

按钮是用得最多的控件，在前面的示例中已经用到了。在 Android 平台中，按钮是通过 Button 来实现的，Button 是一种按钮控件，继承了 TextView，用户能够在该类控件上单击，按钮会触发一个 OnClick 事件。

2.3.1 Button 基本用法

Button 使用起来比较容易，可以通过指定 android:background 属性为按钮增加背景颜色或背景图片，如果将背景图片设置为不规则的，则可以开发出各种不规则形状的按钮。

如果只是使用普通的背景颜色或背景图片，那么这些背景是固定的，不会随着用户的动作而改变。如果需要让按钮的背景颜色、背景图片随用户动作动态改变，则可以考虑使用自定义 Drawable 对象来实现，该部分内容会在高级开发部分进行详细讲解。

接下来通过一个简单的示例程序来学习 Button 的常见用法。具体代码如下。
activit_main.xml：

```xml
<LinearLayout xmlns:android="http://schemas.android.com/apk/res/android"
    xmlns:tools="http://schemas.android.com/tools"
    android:layout_width="match_parent"
    android:layout_height="match_parent"
    android:orientation="vertical"
    android:padding="10dip"
    tools:context="com.hp.main.MainActivity" >
    <!-- 普通文字按钮 -->
    <Button
        android:layout_width="wrap_content"
        android:layout_height="wrap_content"
        android:text="普通按钮"
        android:textSize="16sp" />
    <!-- 图片按钮-->
    <Button
        android:layout_width="60dip"
        android:layout_height="60dip"
        android:background="@drawable/start"   />
    <!-- 带文字的图片按钮-->
    <Button
        android:layout_width="80dp"
        android:layout_height="80dp"
        android:background="@drawable/forward"
        android:textSize="18sp"
        android:text="开始"/>
</LinearLayout>
```

上界面布局中的第一个按钮是一个普通按钮；第二个按钮通过 background 属性配置了背景图片，因此该按钮将会显示为背景图片形状的按钮；第三个按钮综合了文字显示和背景图片，因此该按钮将会显示为背景图片上带文字的按钮。然后修改 MainActivity.java 文件中加载的布局文件为新建的 activit_main.xml 文件。运行程序，可以看到如图 2-9 所示的界面效果。

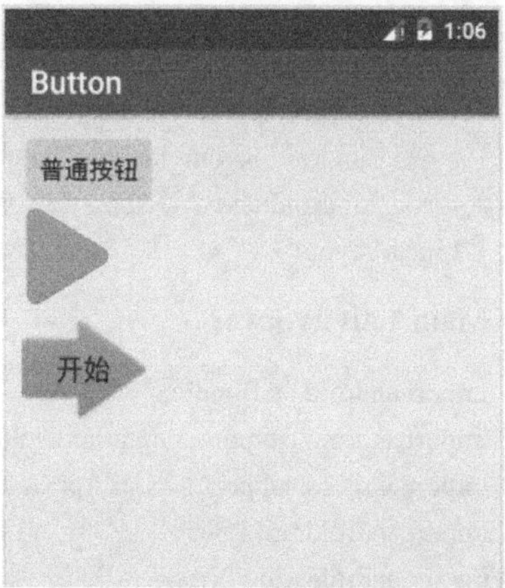

图 2-9

2.3.2 Button 单击事件

通过上面的示例，大体知道如何创建 Button，那么接下来通过一个综合示例来继续学习如何使用 Button 和它的单击事件处理。具体代码如下。

Activity_main.xml:

```xml
<LinearLayout xmlns:android="http://schemas.android.com/apk/res/android"
    xmlns:tools="http://schemas.android.com/tools"
    android:layout_width="match_parent"
    android:layout_height="match_parent"
    android:orientation="vertical"
    android:padding="10dip"
    tools:context="com.hp.main.MainActivity" >
    <TextView
        android:layout_width="match_parent"
        android:layout_height="wrap_content"
        android:text="用户名:"
        android:textSize="16sp"/>
    <EditText
        android:id="@+id/name_et"
        android:layout_width="match_parent"
        android:layout_height="wrap_content"
        android:hint="请输入用户名" />
    <TextView
        android:layout_width="match_parent"
        android:layout_height="wrap_content"
        android:text="密码:"
        android:textSize="16sp"/>
    <EditText
        android:id="@+id/pwd_et"
        android:layout_width="match_parent"
        android:layout_height="wrap_content"
        android:hint="请输入密码"
        android:inputType="textPassword"/>
    <Button
        android:id="@+id/login_btn"
        android:layout_width="match_parent"
        android:layout_height="wrap_content"
        android:text="登录"/>
</LinearLayout>
```

MainActivity.java:

```java
import android.os.Bundle;
import android.support.annotation.Nullable;
import android.support.v7.app.AppCompatActivity;
import android.util.Log;
import android.view.View;
import android.widget.Button;
import android.widget.EditText;

import com.hp.main.R;
```

```java
public class MainActivity extends AppCompatActivity{
    private EditText mNameEt; // 用户名输入框
    private EditText mPasswordEt; // 密码输入框
    private Button mLoginBtn; // 登录按钮

    @Override
    protected void onCreate(@Nullable Bundle savedInstanceState) {
        super.onCreate(savedInstanceState);
        setContentView(R.layout.activity_button2);
        // 获取界面组件
        mNameEt = (EditText) findViewById(R.id.name_et);
        mPasswordEt = (EditText) findViewById(R.id.pwd_et);
        mLoginBtn = (Button) findViewById(R.id.login_btn);
        // 为登录按钮绑定单击事件
        mLoginBtn.setOnClickListener(new View.OnClickListener() {
            @Override
            public void onClick(View view) {
                // 获取用户输入的用户名和密码
                String name = mNameEt.getText().toString();
                String password = mPasswordEt.getText().toString();
                Log.d("MainActivity","用户名: " + name);
                Log.d("MainActivity","密码: " + password);
            }
        });
    }
}
```

其中，若得到 Button 和 EditText 对象，则代码如下。

```java
// 获取界面组件
mNameEt = (EditText) findViewById(R.id.name_et);
mPasswordEt = (EditText) findViewById(R.id.pwd_et);
mLoginBtn = (Button) findViewById(R.id.login_btn);
```

因此，按钮的单击事件代码具体如下。

```java
// 为登录按钮绑定单击事件
        mLoginBtn.setOnClickListener(new View.OnClickListener() {
            @Override
            public void onClick(View view) {
                // 获取用户输入的用户名和密码
                String name = mNameEt.getText().toString();
                String password = mPasswordEt.getText().toString();
                Log.d("MainActivity","用户名: " + name);
                Log.d("MainActivity","密码: " + password);
            }
        });
```

上面的代码采用匿名内部类方式为登录按钮绑定单击事件监听器，运行程序，分别在用户名输入框和密码输入框中输入相应信息，再单击"登录"按钮，可以看到如图2-10所示的界面效果。

```
D/MainActivity: 用户名: tom
D/MainActivity: 密码: 123
```

图 2-10

其实在编码过程中，按钮事件的处理还有其他的方法，第二种就是内部类的方法，具体如下。

```
@Override
protected void onCreate(Bundle savedInstanceState) {
    super.onCreate(savedInstanceState);
    setContentView(R.layout.activity_main);
    btn1.setOnClickListener(new MyClickListener());
}
class MyClickListener implements OnClickListener {
    @Override
    public void onClick(View v) {
        switch (v.getId()) {
            case R.id.btn1:
                //处理逻辑代码
                break;
        }
    }
}
```

还有第三种写法：Activity 实现 View.OnClickListener 接口。

```
public class MainActivity extends Activity implements OnClickListener
```

那么，Activity 就需要实现 onClick()方法，代码如下。

```
//方法：按钮的单击事件
@Override
public void onClick(View v) {
    switch (v.getId()) {
        case R.id.btn1:
            //处理逻辑代码
            break;
        default:
            break;
    }
}
```

2.4 提示框(Toast)

Toast 是 Android 中用来显示信息的一种机制，与 Dialog 不一样的是，Toast 是没有焦点的，而且 Toast 显示的时间有限，经过一定的时间就会自动消失。Toast 类的使用相当简单，而且用途很多，例如，当用户在输入框中输入内容时，可以提示用户输入的内容是否合法；当操作 SQLite 数据库时，提示用户操作的结果。那么 Toast 如何用呢？下面用一个实例来看一看如何使用 Toast。

首先在下面的例子中用到同一个 activity_main.xml，代码如下。

```xml
<LinearLayout xmlns:android="http://schemas.android.com/apk/res/android"
    xmlns:tools="http://schemas.android.com/tools"
    android:layout_width="match_parent"
    android:layout_height="match_parent"
    android:orientation="vertical"
    android:padding="10dip"
    tools:context="com.hp.main.MainActivity" >
    <Button
        android:id="@+id/testBtn"
        android:layout_width="match_parent"
        android:layout_height="wrap_content"
        android:text="测试"/>
</LinearLayout>
```

2.4.1 默认效果

MainActivity.java：

```java
public class MainActivity extends AppCompatActivity{
    private Button testBtn;
    @Override
    protected void onCreate(@Nullable Bundle savedInstanceState) {
        super.onCreate(savedInstanceState);
        setContentView(R.layout.activity_button2);
        testBtn = (Button) findViewById(R.id.testBtn);
        testBtn.setOnClickListener(new View.OnClickListener() {
            @Override
            public void onClick(View view) {
                Toast.makeText(MainActivity.this,"默认 Toast 样式",Toast.LENGTH_LONG).show();
            }
        });
    }
}
```

效果如图 2-11 所示。

使用 Android 高级技术开发 APP

图 2-11

2.4.2 自定义显示位置效果

MainActivity.java：

```
public class MainActivity extends AppCompatActivity{
    private Button testBtn;
    @Override
    protected void onCreate(@Nullable Bundle savedInstanceState) {
        super.onCreate(savedInstanceState);
        setContentView(R.layout.activity_button2);
        // 获取界面组件
        testBtn = (Button) findViewById(R.id.testBtn);
        // 为按钮绑定单击事件
        testBtn.setOnClickListener(new View.OnClickListener() {
            @Override
            public void onClick(View view) {
                Toast toast = Toast.makeText(MainActivity.this,
                        "自定义位置 Toast", Toast.LENGTH_LONG);
                toast.setGravity(Gravity.TOP, 0, 0);
                toast.show();
            }
        });
    }
}
```

效果如图 2-12 所示。

图 2-12

2.4.3 带图片效果

MainActivity.java：

```java
public class MainActivity extends AppCompatActivity{
    private Button testBtn;
    @Override
    protected void onCreate(@Nullable Bundle savedInstanceState) {
        super.onCreate(savedInstanceState);
        setContentView(R.layout.activity_button2);
        // 获取界面组件
        testBtn = (Button) findViewById(R.id.testBtn);
        // 为按钮绑定单击事件
        testBtn.setOnClickListener(new View.OnClickListener() {
            @Override
            public void onClick(View view) {
                Toast toast = Toast.makeText((MainActivity.this,
                        "带图片的 Toast", Toast.LENGTH_LONG);
                toast.setGravity(Gravity.CENTER, 0, 0);
                LinearLayout toastView = (LinearLayout) toast.getView();
                ImageView imageCodeProject = new ImageView(getApplicationContext());
                imageCodeProject.setImageResource
                        (R.drawable.icon);
                toastView.addView
                        (imageCodeProject, 0);
                toast.show();
            }
        });
    }
}
```

效果如图 2-13 所示。

图 2-13

2.5 ImageView

在 Web 开发时，HTML 中对图片的操作就是提供一个<img>标签，通过该标签的 src 属性来制订图片资源的地址，从而在页面中显示一个图片。在 Android 中，使用 ImageView 来显示图片。

ImageView 是用于在界面上展示图片的一个控件，通过它可以让我们的程序界面变得更加丰富多彩。学习 ImageView 控件需要提前准备好一些图片，在 drawable 文件夹下已经有一张 ic_launcher.png 图片，我们就先在界面上展示这张图片，修改 activity_main.xml，代码如下所示：

```xml
<?xml version="1.0" encoding="utf-8"?>
<LinearLayout xmlns:android="http://schemas.android.com/apk/res/android"
    xmlns:tools="http://schemas.android.com/tools"
    android:layout_width="match_parent"
    android:layout_height="match_parent"
    android:orientation="vertical"
    android:padding="10dip"
    tools:context="com.hp.main.MainActivity" >

    <ImageView
        android:layout_width="wrap_content"
        android:layout_height="wrap_content"
        android:src="@drawable/ic_launcher"/>
</LinearLayout>
```

可以看到，这里使用 android:src 属性给 ImageView 指定了一张图片，并且由于图片的宽和高都是未知的，所以将 ImageView 的宽和高都设定为 wrap_content，这样保证了不管图片的尺寸是多少都可以完整地展示出来。运行程序，效果如图 2-14 所示。

图 2-14

关于 ImageView 控件，有如下一些常用属性。

- android:adjustViewBounds：设置 ImageView 是否调整自己的边界来保持所显示图片的长宽比。
- android:maxHeight：设置 ImageView 的最大高度。
- android:maxWidth：设置 ImageView 的最大宽度。
- android:scaleType：设置所显示的图片如何缩放或移动以适应 ImageView 的大小。
- android:src：设置 ImageView 所显示的 Drawable 对象的 ID。

对于 android:scaleType 属性，因为关系到图像在 ImageView 中的显示效果，所以有如下属性值可以选择。

- matrix：使用 matrix 方式进行缩放。
- fitXY：横向、纵向独立缩放，以适应该 ImageView。
- fitStart：保持纵横比缩放图片，并且将图片放在 ImageView 的左上角。
- fitCenter：保持纵横比缩放图片，缩放完成后将图片放在 ImageView 的中央。
- fitEnd：保持纵横比缩放图片，缩放完成后将图片放在 ImageView 的右下角。
- center：把图片放在 ImageView 的中央，但是不进行任何缩放。
- centerCrop：保持纵横比缩放图片，以使图片能完全覆盖 ImageView。
- centerInside：保持纵横比缩放图片，以使得 ImageView 能完全显示该图片。

2.6 ImageButton

ImageButton 是一个可以被用户单击的图片按钮，默认情况下，ImageButton 看起来像一个普通的按钮。按钮的图片可以通过 XML 元素的 android:src 属性或 setImageResource(int) 方法指定。要删除按钮的背景，可以定义自己的背景图片或设置背景为透明。

修改 activity_main.xml，代码如下所示：

```xml
<?xml version="1.0" encoding="utf-8"?>
<LinearLayout xmlns:android="http://schemas.android.com/apk/res/android"
    xmlns:tools="http://schemas.android.com/tools"
    android:layout_width="match_parent"
    android:layout_height="match_parent"
    android:orientation="vertical"
    android:padding="10dip"
    tools:context="com.hp.main.MainActivity" >

    <ImageButton
        android:layout_width="wrap_content"
        android:layout_height="wrap_content"
        android:padding="0dip"
        android:background="#00000000"
        android:src="@drawable/download"/>
</LinearLayout>
```

在上面的布局文件中，设置内边距 padding 的属性值为 0dip，将 ImageButton 的灰色边框去掉，同时，这里也设置了 background 的属性为#00000000，将 ImageButton 的灰色背景去掉。运行程序，效果如图 2-15 所示。

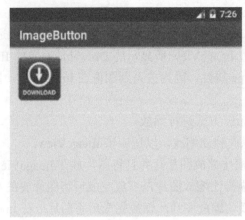

图 2-15

ImageButton 按钮的事件处理方式和事件处理过程同 Button 按钮，此处不再赘述。

2.7 日期和时间框(DatePicker、TimePicker)

日期和时间是任何手机都会有的功能，Android 系统也不例外。Android 平台采用 DatePicker 实现日期，采用 TimerPicker 实现时间，如图 2-16 所示。

图 2-16

Activity_main.xml：

```xml
<LinearLayout xmlns:android="http://schemas.android.com/apk/res/android"
    xmlns:tools="http://schemas.android.com/tools"
    android:layout_width="match_parent"
    android:layout_height="match_parent"
    android:orientation="vertical"
    android:padding="10dip"
    tools:context="com.hp.ui.TextViewActivity" >

    <Button
        android:id="@+id/dateBtn"
        android:layout_width="100dip"
        android:layout_height="wrap_content"
        android:text="选择日期" />
    <Button
        android:id="@+id/timeBtn"
        android:layout_width="100dip"
        android:layout_height="wrap_content"
        android:text="选择时间" />

</LinearLayout>
```

MainActivity.java：

```java
import android.app.DatePickerDialog;
import android.app.TimePickerDialog;
import android.os.Bundle;
import android.support.annotation.Nullable;
import android.support.v7.app.AppCompatActivity;
import android.view.View;
import android.widget.Button;
import android.widget.DatePicker;
import android.widget.TimePicker;
import android.widget.Toast;

import com.hp.main.R;

import java.util.Calendar;

public class MainActivity extends AppCompatActivity{
    private Button dateBtn;
    private Button timeBtn;
    @Override
    protected void onCreate(@Nullable Bundle savedInstanceState) {
        super.onCreate(savedInstanceState);
        setContentView(R.layout.activity_main);
        dateBtn = (Button) findViewById(R.id.dateBtn);
        timeBtn = (Button) findViewById(R.id.timeBtn);
```

```java
dateBtn.setOnClickListener(new View.OnClickListener() {
    @Override
    public void onClick(View view) {
        Calendar c = Calendar.getInstance();
        DatePickerDialog datePickerDialog = new DatePickerDialog(MainActivity.this, new
        DatePickerDialog.OnDateSetListener() {
            @Override
            public void onDateSet(DatePicker view, int year, int monthOfYear,
            int dayOfMonth){

                Toast.makeText(MainActivity.this, year+"-"+ (monthOfYear+1) + "-" +
                dayOfMonth, Toast.LENGTH_SHORT).show();

            }
        },c.get(Calendar.YEAR),c.get(Calendar.MONTH),c.get(Calendar.DAY_OF_MONTH));
        //单击其他部分消失
        datePickerDialog.setCancelable(true);
        datePickerDialog.show();
    }
});
timeBtn.setOnClickListener(new View.OnClickListener() {
    @Override
    public void onClick(View v) {
        Calendar c = Calendar.getInstance();
        TimePickerDialog tp = new TimePickerDialog(MainActivity.this,
        new TimePickerDialog.OnTimeSetListener() {
            @Override
            public void onTimeSet(TimePicker view, int hourOfDay, int minute) {

                Toast.makeText(MainActivity.this, hourOfDay+":"+minute ,
                Toast.LENGTH_SHORT).show();
            }
        },c.get(Calendar.HOUR),c.get(Calendar.MINUTE),true);
        tp.setCancelable(false);
        tp.show();
    }
});
    }
}
```

【单元小结】

- Android 的布局。
- Android 常用控件的使用。
- 掌握如何在配置文件中对控件进行描述。

【单元自测】

1. 以下不属于 Activity 状态的是(   )。
   A. Active   B. Paused   C. Stopped   D. Exit
2. 下面关于 res/raw 目录的说法正确的是(   )。
   A. 这里的文件原封不动地存储到设备上，不会转换为二进制的格式
   B. 这里的文件原封不动地存储到设备上，会转换为二进制的格式
   C. 这里的文件最终以二进制的格式存储到指定的包中
   D. 这里的文件最终不会以二进制的格式存储到指定的包中
3. 下列退出 Activity 的方法错误的是(   )。
   A. finish()   B. 抛异常强制退出
   C. System.exit()   D. onStop()
4. Android 项目工程下的 assets 目录的作用是(   )。
   A. 放置应用到的图片资源
   B. 主要放置多媒体等数据文件
   C. 放置字符串、颜色、数组等常量数据
   D. 放置一些与 UI 相应的布局文件，都是 XML 文件
5. Android 平台采用(   )来实现日期，用(   )来实现时间。
   A. DatePicker   B. TimerPicker   C. Date   D. Timer

【上机实战】

上机目标

- 熟练掌握 Android 常用控件的开发。
- 使用 Android 中的常用控件开发注册、登录页面。

上机练习

练习：Android 常用控件的使用

【问题描述】
使用 Android 常用控件和 Java 与后台进行数据动态交互。

【问题分析】
(1) 使用 XML 完成表示层的工作，由 Java 完成后台数据操作。
(2) 在开发过程中，可以采取先确定 activity_main.xml 中共有多少个控件，然后确定布局方式、线性布局，经过测试无误后，再开发底层。

【参考步骤】

Android 注册界面如图 2-17 所示。

当单击"注册"按钮时，显示如图 2-18 所示。

当单击"取消"按钮时，显示如图 2-19 所示。

图 2-17　　　　　　　图 2-18　　　　　　　图 2-19

具体代码如下。

activity_main.xml：

```
<LinearLayout xmlns:android="http://schemas.android.com/apk/res/android"
    xmlns:tools="http://schemas.android.com/tools"
    android:layout_width="match_parent"
    android:layout_height="match_parent"
    android:orientation="vertical"
    android:padding="10dip"
    tools:context="com.hp.main.MainActivity" >
    <LinearLayout
        android:layout_width="fill_parent"
        android:layout_height="wrap_content" >
        <TextView
            android:layout_width="wrap_content"
            android:layout_height="wrap_content"
            android:text="用户名：" />
        <EditText
            android:id="@+id/txt1"
            android:layout_width="200dip"
            android:layout_height="wrap_content"
            android:hint="请输入用户名"
            android:textSize="14px" />
    </LinearLayout>
    <LinearLayout
        android:layout_width="fill_parent"
        android:layout_height="wrap_content" >
        <TextView
```

```xml
                android:layout_width="wrap_content"
                android:layout_height="wrap_content"
                android:text="密码： " />
            <EditText
                android:id="@+id/txt2"
                android:layout_width="200dip"
                android:layout_height="wrap_content"
                android:hint="请输入密码"
                android:password="true"
                android:textSize="14px" />
        </LinearLayout>
        <LinearLayout
            android:layout_width="fill_parent"
            android:layout_height="wrap_content"
            android:gravity="center"
            android:orientation="horizontal" >
            <Button
                android:id="@+id/btok"
                android:layout_width="100dip"
                android:layout_height="wrap_content"
                android:text="注册" />
            <Button
                android:id="@+id/btres"
                android:layout_width="100dip"
                android:layout_height="wrap_content"
                android:text="取消" />
        </LinearLayout>
</LinearLayout>
```

MainActivity.java：

```java
import android.os.Bundle;
import android.support.annotation.Nullable;
import android.support.v7.app.AppCompatActivity;
import android.view.Gravity;
import android.view.View;
import android.widget.Button;
import android.widget.EditText;
import android.widget.Toast;

import com.hp.main.R;

public class MainActivity extends AppCompatActivity{
    private EditText tname;
    private EditText tpsw;
    private Button btok;
    private Button btres;
    @Override
```

```java
protected void onCreate(@Nullable Bundle savedInstanceState) {
    super.onCreate(savedInstanceState);
    setContentView(R.layout.activity_main);
    tname=(EditText) this.findViewById(R.id.txt1);
    tpsw=(EditText) this.findViewById(R.id.txt2);
    btok=(Button) this.findViewById(R.id.btok);
    btres=(Button) this.findViewById(R.id.btres);
    btok.setOnClickListener(new View.OnClickListener() {
        @Override
        public void onClick(View v) {
            String name=tname.getText().toString();
            String psw=tpsw.getText().toString();
            String str="你输入的账号是:"+name+",你输入的密码是:"+psw;
            myPlay(str);
        }
    });
    btres.setOnClickListener(new View.OnClickListener() {
        @Override
        public void onClick(View v) {
            tname.setText("");
            tpsw.setText("");
        }
    });
}
/**
 * 提示
 */
public void myPlay(String str)
{
    Toast t=Toast.makeText(this,str,Toast.LENGTH_SHORT);
    t.setGravity(Gravity.TOP,0,200);
    t.show();
}
}
```

【拓展作业】

1. 请在 Android 平台上完成学生信息注册页面。
2. 介绍一下 Android 常用控件的共同特点。
3. 尝试将所有的常用控件放在一个 Activity 中实现。

单元三

Activity 和 Intent

课程目标

- ▶ Activity 介绍
- ▶ Activity 生命周期
- ▶ Intent 介绍和使用

 简 介

在 Android 中，多数情况下每个程序都是在各自独立的 Linux 进程中运行的。当一个程序或其某些部分被请求时，它的进程就"出生"了，当这个程序没有必要再运行下去且系统需要回收这个进程的内存用于其他程序时，这个进程就"死亡"了。可以看出，Android 程序的生命周期是由系统控制而非程序自身直接控制。一个 Android 程序的进程是何时被系统结束的呢？通俗地说，一个即将被系统关闭的程序是系统在内存不足(lowmemory)时，根据"重要性层次"选出来的"牺牲品"。一个进程的重要性是根据其中运行的部件和部件的状态决定的。各种进程按照重要性从高到低排列如下：前台进程、可见进程、服务进程、后台进程、空进程。从某种意义上讲，垃圾收集机制把程序员从"内存管理噩梦"中解放出来，而 Android 的进程生命周期管理机制把用户从"任务管理噩梦"中解放出来。Android 使用 Java 作为应用程序 API，并且结合其独特的生命周期管理机制同时为开发者和使用者提供最大程度的便利。本单元重点介绍 Android 生命周期和 Intent 的使用。

3.1 Activity 的介绍

3.1.1 Activity 的概念

Activity(活动)是一种包含用户界面的组件，主要用于和用户进行交互(可以简单理解为 APP 中的一个页面)。一个应用程序中可以包含多个活动，也可以不包含活动。

通过前面两个单元的介绍，我们已经进一步对活动有了初步的认识。本单元将对活动进行详细的介绍。

3.1.2 活动的基本用法

由于 Android Studio 在一个工作区间内只允许打开一个项目，因此首先需要将当前的项目关闭，单击导航栏 File→Close Project，然后再新建一个 Android 项目，但需要做一点改动，即不再选择 Empty Activity 选项，而是选择 Add No Activity 选项，因为这次我们准备手动创建活动，如图 3-1 所示。单击 Finish 按钮，等待项目构建完成。

项目创建成功后，目录 app/src/main/java/com.example.luyong.activitytest 中就没有 MainActivity 类了，如图 3-2 所示。

单元三 Activity和Intent

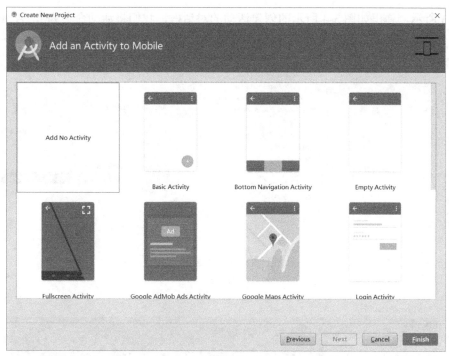

图 3-1

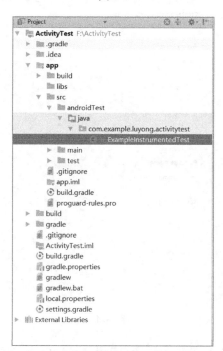

图 3-2

3.1.3 创建活动

在 com.example.luyong.activitytest 上右击，选择 New→Activity→Empty Activity 选项，

如图 3-3 所示，弹出如图 3-4 所示的创建活动的对话框，我们将活动命名为 MyActivity，并且不勾选 Generate Layout File 和 Launcher Activity 两个选项。

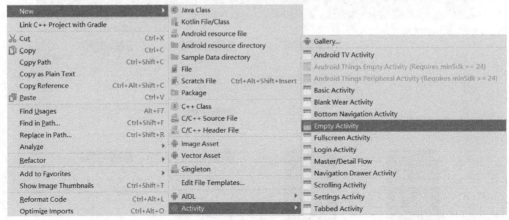

图 3-3

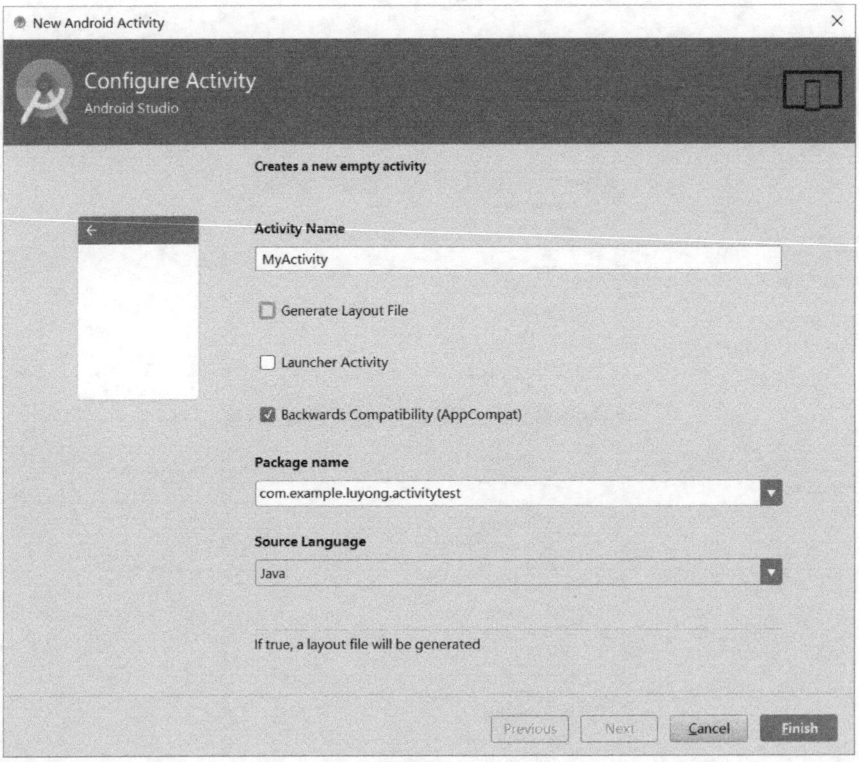

图 3-4

勾选 Generate Layout File 选项表示会自动为 MyActivity 创建一个对应的布局文件；勾选 Launcher Activity 选项表示会自动将 MyActivity 设置为当前项目的主活动，这里由于我们是第一次手动创建活动，因此自动生成的选项暂时都不要勾选，下面将会一个个手动来完成。勾选 Backwards Compatibility 选项表示会为项目启用向下兼容的模式，需勾选该选项。单击 Finish 按钮完成创建。

项目中的任何活动都应该重写 Activity 的 onCreate()方法，而目前 MyActivity 中已经重写了这个方法，这是由 Android Studio 自动帮我们完成的。

3.1.4 创建布局

Android 程序的设计讲究逻辑和视图分离，而布局就是用来显示界面内容的，因此我们现在就来手动创建一个布局文件。

右击 app/src/main/res 目录，选择 New→Directory 选项，如图 3-5 所示，创建一个文件夹 layout 用于存储我们的布局文件，如图 3-6 所示。

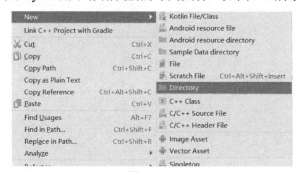

图 3-5

图 3-6

右击 layout 目录，选择 New→XML→Layout XML file 选项，如图 3-7 所示，弹出一个新建布局资源文件的窗口，我们将这个布局文件命名为 my_layout，根元素默认选择为 LinearLayout，如图 3-8 所示。

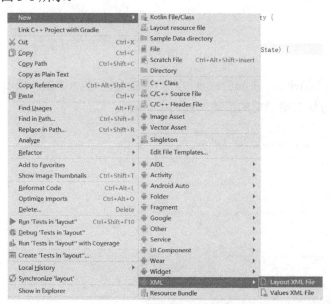

图 3-7

由于刚才在创建布局文件时选择了 LinearLayout 作为根元素，因此现在的布局已经有一个 LinearLayout 元素了，现在我们添加一个按钮，如图 3-9 所示。

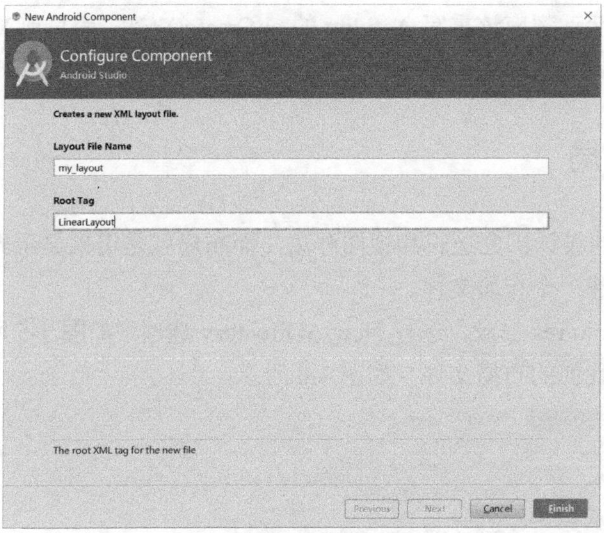

图 3-8

```
<?xml version="1.0" encoding="utf-8"?>
<LinearLayout xmlns:android="http://schemas.android.com/apk/res/android"
    android:layout_width="match_parent"
    android:layout_height="match_parent">
    <Button
        android:id="@+id/button_1"
        android:layout_width="match_parent"
        android:layout_height="wrap_content"
        android:text="Button1"/>


</LinearLayout>
```

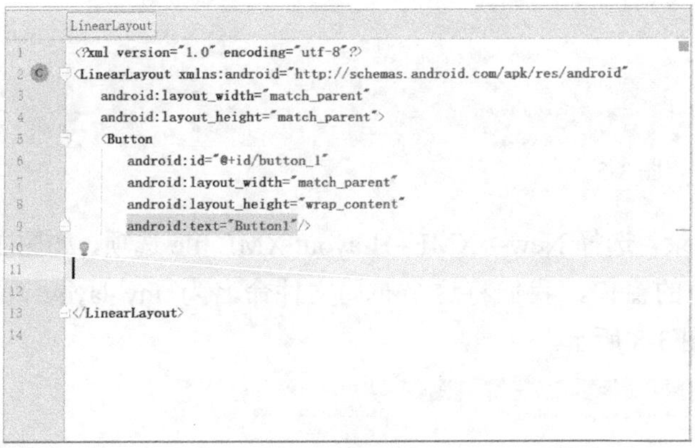

图 3-9

重新回到MyActivity，在onCreate()方法中加入代码，如图3-10所示。这里调用了setContentView()方法来给当前的活动加载一个布局。

```
package com.example.luyong.activitytest;

import ...

public class MyActivity extends AppCompatActivity {

    @Override
    protected void onCreate(Bundle savedInstanceState) {
        super.onCreate(savedInstanceState);
        setContentView(R.layout.my_layout);
    }
}
```

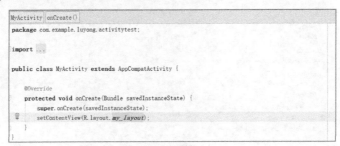

图 3-10

3.1.5 AndroidManifest 文件中注册 Activity

所有的活动都要在 AndroidManifest.xml 中进行注册才能生效，而实际上 FirstActivity

已经在 AndroidManifest.xml 中注册过了，打开 app/src/main/AndroidManifest.xml 文件，代码如图 3-11 所示。

```
<?xml version="1.0" encoding="utf-8"?>
<manifest xmlns:android="http://schemas.android.com/apk/res/android"
    package="com.example.luyong.activitytest">

    <application
        android:allowBackup="true"
        android:icon="@mipmap/ic_launcher"
        android:label="ActivityTest"
        android:roundIcon="@mipmap/ic_launcher_round"
        android:supportsRtl="true"
        android:theme="@style/AppTheme">
        <activity android:name=".MyActivity"></activity>
    </application>

</manifest>
```

图 3-11

可以看到，活动的注册声明要放在<application>标签内，这里通过<activity>标签对活动进行注册。不过这些任务 Android Studio 已经帮我们做过了。

在<activity>标签中我们使用了android:name来指定具体注册哪一个活动，这里填入的.MyActivity是com.example.luyong.activitytest.MyActivity的缩写。由于在最外层的<manifest>标签中已经通过package属性指定了程序的包名是com.example.luyong.activitytest，因此在注册活动时这一部分可以省略，直接使用.FirstActivity即可。

不过，这样仅是注册了活动，程序仍然是不能运行的，因为还没有为程序配置主活动，当程序运行起来时，不知道要先启动哪个活动。在<activity>标签的内部加入 intent-filter>标签，并在该标签中添加<actionandroid:name="android.intent.action. MAIN"/>,<category android name="android.intent.category.LAUNCHER"/>两句声明即可。 除此之外，我们还可以使用 android:label 指定活动中标题栏的内容，标题栏显示在活动的最顶部。修改后的AndroidManifest.xml 文件，代码如图 3-12 所示，运行程序，结果如图 3-13 所示。

```
<?xml version="1.0" encoding="utf-8"?>
<manifest xmlns:android="http://schemas.android.com/apk/res/android"
    package="com.example.luyong.activitytest">

    <application
        android:allowBackup="true"
        android:icon="@mipmap/ic_launcher"
        android:label="ActivityTest"
        android:roundIcon="@mipmap/ic_launcher_round"
        android:supportsRtl="true"
        android:theme="@style/AppTheme">
        <activity android:name=".MyActivity">
            <intent-filter>
                <action android:name="android.intent.action.MAIN" />
                <category  android:name="android.intent.category.LAUNCHER"  />
            </intent-filter>
        </activity>
    </application>

</manifest>
```

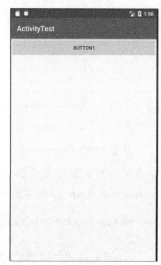

图 3-12

图 3-13

3.2 Activity 的生命周期

发布程序到手机上之后，当双击该项目应用的图标时，系统会将这个单击事件包装成一个 Intent，该意图被传递给应用，在应用的功能清单文件中寻找与该意图匹配的意图过滤器，如果匹配成功，则找到相匹配的意图过滤器所在的 Activity 元素，再根据<activity>元素的 name 属性来寻找其对应的 Activity 类；接着 Android 操作系统创建该 Activity 类的实例对象，对象创建完成之后，会执行到该类的 onCreate()方法，该 OnCreate()方法通过重写其父类 Activity 的 OnCreate()方法而实现。onCreate()方法用来初始化 Activity 实例对象。Android Activity 的生命周期如图 3-14 所示。

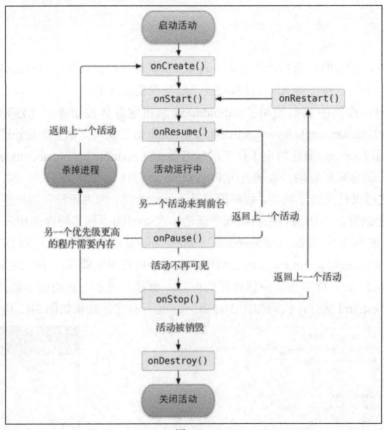

图 3-14

从新建的 Activity 中可以看出 Activity 继承了 ApplicationContext 类，我们可以重写以下方法，为了更好地了解 Android Activity 的生命周期，做如下测试，即可在短时间之内明白 Android Activity 的生命周期，测试的每个方法中都使用 Log 打印输出当前的方法启动。

```
package com.example.luyong.activitytest;

import android.support.v7.app.AppCompatActivity;
import android.os.Bundle;public class MainActivity extends Activity
{
```

```java
/***
 * 主要用 Log 打印
 */
@Override
protected void onCreate(Bundle savedInstanceState)
{
    super.onCreate(savedInstanceState);
    setContentView(R.layout.my_layout);
    Log.e("当前应用程序","onCreate 启动了！");
}
@Override
protected void onStart()
{
    // TODO Auto-generated method stub
    super.onStart();
    Log.e("当前应用程序", "onStart 启动了");
}
@Override
protected void onRestart()
{
    // TODO Auto-generated method stub
    super.onRestart();
    Log.e("当前应用程序","onRestart 启动了！");
}
@Override
protected void onResume()
{
    // TODO Auto-generated method stub
    super.onResume();
    Log.e("当前应用程序","onResume 启动了");
}
@Override
protected void onPause()
{
    // TODO Auto-generated method stub
    super.onPause();
    Log.e("当前应用程序","onPause 启动了");
}
@Override
protected void onStop()
{
    // TODO Auto-generated method stub
    super.onStop();
    Log.e("当前应用程序","onStop 启动了");
}
@Override
protected void onDestroy()
```

```
        {
            // TODO Auto-generated method stub
            super.onDestroy();
            Log.e("当前应用程序","onDestroy 启动了");
        }
        @Override
}
```

运行该应用程序，查看 LogCat 窗口，可以看到它的执行过程为 onCreate()→onStart()→onResume()，如图 3-15 所示。

```
04-17 02:40:05.752  4024-4024/com.example.luyong.activitytest I/当前应用程序: onCreate启动了
04-17 02:40:05.754  4024-4024/com.example.luyong.activitytest I/当前应用程序: onStart启动了
04-17 02:40:05.755  4024-4024/com.example.luyong.activitytest I/当前应用程序: onResume启动了
```

图 3-15

单击"返回"按钮时该应用程序结束。这里调用的应用程序是 onPause()→onStop()→onDestroy()，查看 LogCat 窗口，可以看到如图 3-16 所示的结果。

```
04-17 02:41:50.581  4024-4024/com.example.luyong.activitytest I/当前应用程序: onPause启动了
04-17 02:41:50.601  4024-4024/com.example.luyong.activitytest I/当前应用程序: onStop启动了
04-17 02:41:50.601  4024-4024/com.example.luyong.activitytest I/当前应用程序: onDestroy启动了
```

图 3-16

当打开应用程序时，如正在浏览 NBA 新闻，看到一半时，突然想听歌，这时我们会选择按 HOME 键退出，然后打开音乐应用程序。按 HOME 键时，Activity 先后执行了 onPause()、onStop()两个方法，而应用程序并没有被销毁，如图 3-17 所示。

```
04-17 02:43:26.427  4024-4024/com.example.luyong.activitytest I/当前应用程序: onPause启动了
04-17 02:43:26.435  4024-4024/com.example.luyong.activitytest I/当前应用程序: onStop启动了
```

图 3-17

当我们再次启动MainActivity应用程序时，先后分别执行了onRestart()、onStart()、onResume()三个方法，如图 3-18 所示。

```
04-17 02:44:30.929  4024-4024/com.example.luyong.activitytest I/当前应用程序: onRestart启动了
04-17 02:44:30.931  4024-4024/com.example.luyong.activitytest I/当前应用程序: onStart启动了
04-17 02:44:30.931  4024-4024/com.example.luyong.activitytest I/当前应用程序: onResume启动了
```

图 3-18

回顾上面 Activity 生命周期图，大家就会很熟悉 Activity 的生命周期了。

Activity 类中定义了 7 个回调方法，覆盖了活动生命周期的每一个环节，下面具体介绍。

- onCreate()方法。该方法我们已经看到过很多次了，每个活动中都重写了这个方法，它会在活动第一次被创建的时候调用。我们会在这个方法中完成活动的初始化操作，如加载布局、绑定事件等。

- onStart()方法。该方法在活动由不可见变为可见的时候调用。
- onResume()方法。该方法在活动准备好和用户进行交互的时候调用。此时的活动一定位于返回栈的栈顶，并且处于运行状态。
- onPause()方法。该方法在系统准备去启动或者恢复另一个活动的时候调用。
- onStop()方法。该方法在活动完全不可见的时候调用。它与onPause()方法的主要区别在于，如果启动的新活动是一个对话框式的活动，那么 onPause()方法会得到执行，而 onStop()方法并不会执行。
- onDestroy()方法。该方法在活动被销毁之前调用，之后活动的状态将变为销毁状态。
- onRestart()方法。该方法在活动由停止状态变为运行状态之前调用，也就是活动被重新启动了。

以上 7 个方法中，除了 onRestart()方法外，其他都是两两相对的，从而又可以将活动分为以下 3 种生存期。

- 完整生存期。活动在 onCreate()方法和 onDestroy()方法之间所经历的，就是完整生存期。一个活动会在 onCreate()方法中完成各种初始化操作，而在 onDestroy()方法中完成释放内存的操作。
- 可见生存期。活动在 onStart()方法和 onStop()方法之间所经历的，就是可见生存期。在可见生存期内，活动对于用户总是可见的，即便有可能无法和用户进行交互。我们可以通过这两个方法，合理地管理对用户可见的资源。例如，在 onStart()方法中对资源进行加载，而在 onStop()方法中对资源进行释放，从而保证处于停止状态的活动不会占用过多内存。
- 前台生存期。活动在 onResume()方法和 onPause()方法之间所经历的，就是前台生存期。在前台生存期内，活动总是处于运行状态，此时的活动是可以和用户进行交互的，我们平时看到和接触最多的也就是这个状态下的活动。

3.3 Intent 介绍和使用

当我们有多个活动时，在启动器中单击应用的图标只会进入该应用的主活动，那么怎样才能由主活动跳转到其他活动呢？我们现在就来一起看一看。

现在我们快速地在ActivityTest项目中再创建一个活动并命名为SecondActivity。勾选Generate Layout File选项，Android Studio会自动生成SecondActivity.java和second_layout.xml两个文件。

我们把布局文件改成与 my_layout.xml 一样，但要修改按钮的 ID 和名称，如图 3-19 所示。

此时一个程序就有了两个活动，那么怎么让程序从MyActivity跳转到SecondActivity呢？这就用到了Intent。Intent是Android程序中各组件之间进行交互的一种重要方式，它不仅可以指明当前组件想要执行的动作，还可以在不同组件之间传递数据。Intent一般可被用于启动活动、启动服务及发送广播等。

```xml
<?xml version="1.0" encoding="utf-8"?>
<LinearLayout xmlns:android="http://schemas.android.com/apk/res/android"
    android:layout_width="match_parent"
    android:layout_height="match_parent">
    <Button
        android:id="@+id/button_2"
        android:layout_width="match_parent"
        android:layout_height="wrap_content"
        android:text="Button2"/>

</LinearLayout>
```

图 3-19

Intent 可分为两种：显式 Intent 和隐式 Intent。下面我们具体介绍。

3.3.1 显式 Intent

显式 Intent，即直接指定需要打开的 Activity 对应的类。
在 MyActivity 中给按钮添加一个单击事件，代码如下。

```java
button1.setOnClickListener(new View.OnClickListener() {
    @Override
    public void onClick(View v) {
        Intent intent = new Intent(MyActivity.this,SecondActivity.class);
        startActivity(intent);
    }
});
```

重新运行程序，在 FirstActivity 界面中单击一下按钮，结果如图 3-20 所示。

图 3-20

可以看到，我们已经成功启动 SecondActivity 活动。使用这种方式启动活动时，Intent 的"意图"非常明显，因此称为显式 Intent。

3.3.2 隐式 Intent

相比于显式 Intent，隐式 Intent 则含蓄了许多，它并不明确指出我们想要启动哪一个活动，而是指定了 action 和 category 等信息，然后交由系统去分析这个 Intent，并帮我们找出合适的活动去启动。

关于隐式 Intent 的操作，我们需修改 AndroidManifest.xml 配置文件，即在<activity>标签下配置<intent-filter>，指定当前活动能够响应的 action 和 category，添加代码如下。

```
<activity android:name=".SecondActivity">
<intent-filter>
<action android:name="com.example.luyong.activitytest.ACTION_START"/>
<category android:name="android.intent.category.DEFAULT" />
    </intent-filter>
</activity>
```

在<action>标签中我们指明了当前活动可以响应 com.example.luyong.activitytest.ACTION_START 这个 action，而<category>标签则包含了一些附加信息，更精确地指明了当前活动能够响应的 Intent 中还可能带有的 category。只有<action>和<category>中的内容同时能够匹配上 Intent 中指定的 action 和 category 时，这个活动才能响应该 Intent。

修改 FirstActivity 中按钮的单击事件，代码如下。

```
button1.setOnClickListener(new View.OnClickListener() {
    @Override
    public void onClick(View v) {
        Intent intent = new Intent("com.example.luyong.activitytest.ACTION_START");
        startActivity(intent);
    }
});
```

重新运行程序，在 MyActivity 界面单击一下按钮，即可成功启动 SecondActivity。不同的是，这次我们使用了隐式 Intent 的方式来启动，说明在<activity>标签下配置的 action 和 category 的内容已经生效了！

每个 Intent 中只能指定一个 action，但却能指定多个 category。Intent 中只有一个默认的 category，我们现在再增加一个，修改 FirstActivity 中按钮的单击事件，代码如下。

```
button1.setOnClickListener(new View.OnClickListener() {
    @Override
    public void onClick(View v) {
        Intent intent = new Intent("com.example.luyong.activitytest.ACTION_START");
        intent.addCategory("com.example.luyong.activitytest.MY_CATEGORY");
```

```
        startActivity(intent);
    }
});
```

可以调用 Intent 中的 addCategory()来添加一个 category，这里我们指定了一个自定义的 category 值为 com.example.luyong.activitytest.MY_CATEGORY。

现在我们在<intent-filter>中再添加一个 category 的声明，代码如下所示。

```
<activity android:name=".SecondActivity">
<intent-filter>
<action android:name="com.example.activitytest.ACTION_START" />
    <category android:name="android.intent.category.DEFAULT"/>
    <category android:name="com.example.luyong.activitytest.MY_CATEGORY"/>
</intent-filter>
</activity>
```

运行之后我们发现和显式的效果是一样的，也可以执行跳转。

3.4 Intent 传值和传对象

3.4.1 Intent 传值

Intent 还可以在启动活动的时候传递数据，下面我们一起来看一下。在启动活动时传递数据很简单，Intent 中提供了一系列 putExtra()方法的重载，可以把我们想要传递的数据暂存在 Intent 中，启动另一个活动后，把这些数据再从 Intent 中取出即可。例如，MyActivity 中有一个字符串，现在想把这个字符串传递到 SecondActivity 中，可以编写代码如下。

```
button1.setOnClickListener(new View.OnClickListener() {
    @Override
    public void onClick(View v) {
        String str = "Hello World!";
        Intent intent = new Intent(MyActivity.this,SecondActivity.class);
        intent.putExtra("str", str);
        startActivity(intent);
    }
});
```

这里 putExtra()方法接收两个参数，第一个参数是键，用于后面从 Intent 中取值，第二个参数才是真正要传递的数据。

在 SecondActivity 中将传递的数据取出并打印出来，代码如下。

```
public class SecondActivity extends AppCompatActivity {
    @Override
    protected void onCreate(Bundle savedInstanceState) {
        super.onCreate(savedInstanceState);
        setContentView(R.layout.second_layout);
```

```
            Intent intent = getIntent();
            String str = intent.getStringExtra("str");
            Log.d("message",str);
    }
}
```

首先通过 getIntent() 方法获取用于启动 SecondActivity 的 Intent，然后调用 getStringExtra() 方法，传入相应的键值，就可以得到传递的数据了。

重新运行程序，在 MyActivity 界面中单击一下按钮会跳转到 SecondActivity，查看 logcat 打印信息，如图 3-21 所示。

```
04-17 06:35:24.070 4772-4772/com.example.luyong.activitytest D/message: Hello World!
```
图 3-21

3.4.2　Intent 传对象

Android 为 Intent 提供了两种传递对象参数类型的方法，分别需要使实体类实现 Serializable 接口，我们要知道，传递对象，需要先将对象序列化。

将需要传送的对象所属的实体类实现 Serializable 接口，代码如下。

```
package com.example.luyong.entity;

import java.io.Serializable;
public class UserInfo implements Serializable {
    int userid;
    String username;
    String usersex;

    public int getUserid() {
        return userid;
    }

    public void setUserid(int userid) {
        this.userid = userid;
    }

    public String getUsername() {
        return username;
    }

    public void setUsername(String username) {
        this.username = username;
    }

    public String getUsersex() {
        return usersex;
```

```java
        }
        public void setUsersex(String usersex) {
            this.usersex = usersex;
        }
        @Override
        public String toString() {
            return "UserInfo{" +
                    "userid=" + userid +
                    ", username='" + username + '\'' +
                    ", usersex='" + usersex + '\'' +
                    '}';
        }
    }
```

传递数据的步骤如下。

(1) 在 MyActivity 的单击事件中写入如下代码。

```java
button1.setOnClickListener(new View.OnClickListener() {
    @Override
    public void onClick(View view) {
        Intent intent = new Intent(MyActivity.this,SecondActivity.class);
        UserInfo user=new UserInfo();
        user.setUserid(1);
        user.setUsername("张三");
        user.setUsersex("男");
        intent.putExtra("user", user);
        startActivity(intent);
    }
});
```

(2) 在 SecondActivity 中取出数据并打印输出，代码如下。

```java
protected void onCreate(Bundle savedInstanceState) {
    super.onCreate(savedInstanceState);
    setContentView(R.layout.activity_second);
    Intent intent = getIntent();
    UserInfo user = (UserInfo) intent.getSerializableExtra("user");
    Log.d("message",user.toString());

}
```

(3) 运行项目，查看 LogCat 打印信息，如图 3-22 所示。

```
04-17 07:09:47.470 5369-5369/com.example.luyong.activitytest D/message: UserInfo{userid=1, username='张三', usersex='男'}
```

图 3-22

单元三 Activity和Intent

【单元小结】

- Activity 的介绍。
- Activity 的生命周期。
- 使用 Intent 传递数据。

【单元自测】

1. string.xml 的主要作用是(　　)。
   A. 常量定义文件　　　　　　B. 变量定义文件
   C. 数据库连接文件　　　　　D. 流程控制文件
2. R.java 包含的静态内部类有(　　)。
   A. attr　　　　B. drawable　　　C. layout　　　D. string
3. AndroidManifest.xml 的主要作用是(　　)。
   A. 提供了关于这个应用程序的基本信息
   B. 提供了关于连接数据库的基本信息
   C. 变量定义文件
   D. 常量定义文件
4. Activity 的父类是(　　)。
   A. ApplicationContext　　　　B. AppContext
   C. Application　　　　　　　D. InitialContext
5. Activity 生命周期一共有(　　)个方法。
   A. 4　　　　　B. 5　　　　　C. 6　　　　　D. 7

【上机实战】

上机目标

- 了解 Android Activity 的生命周期。
- 使用实例验证 Android Activity 的生命周期。

上机练习

练习：在 LogCat 下查看 Android 的生命周期

【问题描述】
使用 Android 完成用户注册程序。

使用 Android 高级技术开发 APP

【问题分析】

(1) 使用 Android 的 XML 完成表示层的工作，由 Java 完成后台控制。

(2) 在开发过程中，可以采取先完成页面布局，也就是完成XML的设置。本例中首先编写activity_main.xml，经过一个测试类测试底层无误后，再来开发底层控制，表现层和底层应分别测试。

【参考步骤】

Activity-main.xml:

```xml
<RelativeLayout xmlns:android="http://schemas.android.com/apk/res/android"
    xmlns:tools="http://schemas.android.com/tools"
    android:layout_width="match_parent"
    android:layout_height="match_parent"
    android:paddingBottom="@dimen/activity_vertical_margin"
    android:paddingLeft="@dimen/activity_horizontal_margin"
    android:paddingRight="@dimen/activity_horizontal_margin"
    android:paddingTop="@dimen/activity_vertical_margin"
    tools:context=".MainActivity" >
    <LinearLayout
        android:layout_width="fill_parent"
        android:layout_height="fill_parent"
        android:orientation="vertical" >
        <LinearLayout
            android:layout_width="fill_parent"
            android:layout_height="wrap_content" >
            <TextView
                android:layout_width="wrap_content"
                android:layout_height="wrap_content"
                android:text="@string/myName" />
            <EditText
                android:id="@+id/txt1"
                android:layout_width="200dip"
                android:layout_height="wrap_content"
                android:hint="@string/txtNameMessage"
                android:textSize="14px" />
        </LinearLayout>
        <LinearLayout
            android:layout_width="fill_parent"
            android:layout_height="wrap_content" >
            <Button
                android:id="@+id/btn_ok"
                android:layout_width="100dip"
                android:layout_height="wrap_content"
                android:text="@string/myok" />
        </LinearLayout>
    </LinearLayout>
```

```
</RelativeLayout>
```

String.xml：

```xml
<?xml version="1.0" encoding="utf-8"?>
<resources>
    <string name="app_name">注册</string>
    <string name="action_settings">Settings</string>
    <string name="myName">账号：</string>
    <string name="txtNameMessage">请输入账号</string>
    <string name="myok">确定</string>
</resources>
```

MainActivity.java：

```java
package com.hp.hello;
import android.app.Activity;
import android.os.Bundle;
import android.util.Log;
import android.view.Menu;
public class MainActivity extends Activity
{
private Button btn_ok;
private EditText txt1;
    /***
     * 主要用 Log 打印
     */
    @Override
    protected void onCreate(Bundle savedInstanceState)
    {
        super.onCreate(savedInstanceState);
        setContentView(R.layout.activity_main);
btn_ok = (Button) findViewById(R.id.btn_ok);
txt1 = (EditText) findViewById(R.id.txt1);
System.out.println(txt.getText().toString());
        Log.e("当前应用程序","onCreate 启动了！");
    }
    @Override
    protected void onStart()
    {
        // TODO Auto-generated method stub
        super.onStart();
        Log.e("当前应用程序", "onStart 启动了");
    }
    @Override
    protected void onRestart()
    {
        // TODO Auto-generated method stub
        super.onRestart();
```

```
            Log.e("当前应用程序","onRestart 启动了！");
        }
        @Override
        protected void onResume()
        {
            // TODO Auto-generated method stub
            super.onResume();
            Log.e("当前应用程序","onResume 启动了");
        }
        @Override
        protected void onPause()
        {
            // TODO Auto-generated method stub
            super.onPause();
            Log.e("当前应用程序","onPause 启动了");
        }
        @Override
        protected void onStop()
        {
            // TODO Auto-generated method stub
            super.onStop();
            Log.e("当前应用程序","onStop 启动了");
        }
        @Override
        protected void onDestroy()
        {
            // TODO Auto-generated method stub
            super.onDestroy();
            Log.e("当前应用程序","onDestroy 启动了");
        }
        @Override
        public boolean onCreateOptionsMenu(Menu menu)
        {
            getMenuInflater().inflate(R.menu.main, menu);
            return true;
        }
    }
```

运行程序，并查看控制台情况。

【拓展作业】

1. 请在 Android 平台模拟一个登录功能，要求体现并测试 Android 的生命周期。
2. 请在 MyEclipse 中总结观察 Activity 生命周期的先后顺序。

单元四 Android UI 布局详解

 课程目标

- ▶ Android UI 布局的介绍
- ▶ Android UI 常用六大布局

 简 介

布局(Layout)的概念是针对 Activity 的，Activity 就是布满整个 Android 设备的窗口或者悬浮于其他窗口上的交互界面。一个应用程序通常由多个 Activity 构成，每个需要显示的 Activity 都需要在 AndroidManifest.xml 文件中声明。通常情况下，开发人员可以使用两种方式来创建 UI 组件，一种是使用 XML 方式来配置 UI 组件的相关属性，然后装载这些 UI 组件，这也是最常用的方式。但是有些特殊情况下，需要动态生成 UI 组件，则需要使用第二种方式，即完全使用 Java 代码来创建 UI 组件。本单元将重点介绍用 XML 方式来设计 UI 布局。

4.1 Android UI 布局介绍

在 Android 中，共有 6 种布局方式，分别是 FrameLayout(框架布局)、LinearLayout(线性布局)、AbsoluteLayout(绝对布局)、RelativeLayout(相对布局)、TableLayout(表格布局)和 GridLayout(网格布局)。各布局特点及应用场景介绍如下。

1. FrameLayout 框架布局

布局特点：放入其中的所有元素都被放置在最左上的区域，而且无法为这些元素指定一个确切的位置，下一个子元素会重叠覆盖上一个子元素。

应用场景：适合浏览单张图片。

2. LinearLayout 线性布局

布局特点：主要提供控件水平或者垂直排列的模型，每个子组件都可以垂直或水平的方式线性排布(默认是垂直)。

应用场景：最常用的布局方式。

LinearLayout 中有一个重要的属性 android:layout_weight="1"，其中，weight 在垂直布局时，代表行距，水平布局时，代表列宽；weight 值越大，行距或列宽就越大。

3. AbsoluteLayout 绝对布局

布局特点：采用坐标轴的方式定位组件，左上角是(0, 0)点，往右 x 轴递增，往下 y 轴递增，组件定位属性为 android:layout_x 和 android:layout_y 来确定坐标。

应用场景：准确定位空间位置。

由于 Android 手机的屏幕尺寸、分辨率存在较大差异，使用 AbsoluteLayout 无法兼顾适配问题，所以该布局已经过时。

4. RelativeLayout 相对布局

布局特点：以某一个组件为参照物，来定位下一个组件的位置的布局方式。

应用场景：控件之间存在相应关系(适配性较好，推荐使用)。

5. TableLayout 表格布局

布局特点：类似 HTML 中的 Table，使用 TableRow 来布局，其中 TableRow 代表一行，TableRow 的每一个视图组件代表一个单元格。

应用场景：控件之间存在相应关系。

6. GridLayout 网格布局

GridLayout 布局是 Android 4.0 以后引入的新布局，与 TableLayout(表格布局)类似，不过它功能更多，也更加好用，具体如下。

- 可以自己设置布局中组件的排列方式。
- 可以自定义网格布局的行数和列数。
- 可以直接设置组件位于某行某列。
- 可以设置组件横跨几行或者几列。

以上 6 个布局元素也可以相互嵌套应用，做出更美观、复杂的界面。

4.2 Android UI 常用六大布局

4.2.1 LinearLayout 线性布局详解

LinearLayout 容器中的组件挨个排列，通过控制 android:orientation 属性，可控制各组件是横向排列还是纵向排列。

LinearLayout 的常用 XML 属性、相关方法及说明，如表 4-1 所示。

表 4-1

XML 属性	相关方法	说明
android:gravity	setGravity(int)	设置布局管理器内组件的对齐方式
android:orientation	setOrientation(int)	设置布局管理器内组件的排列方式，可以设置为 horizontal、vertical 两个值之一

LinearLayout 子元素支持的常用 XML 属性及说明，如表 4-2 所示。

表 4-2

XML 属性	说明
android:layout_gravity	指定子元素在 LinearLayout 中的对齐方式
android:layout_weight	指定子元素在 LinearLayout 中所占的权重

LinearLayout 常用的属性值有如下几种。

- android:gravity="center_horizontal"　子控件水平方向居中

- android:gravity="center_vertical" 子控件垂直方向居中
- android:gravity="center" 子控件垂直方向和水平方向居中
- android:gravity= start || end || top || bottom 子控件左对齐 || 右对齐 || 顶部对齐 || 底部对齐
- android:gravity= left || right 子控件左对齐 || 右对齐

这里的 start 和 left 属性、end 和 right 属性需要注意一下，此处是针对中国的情况而言。实际上，它们两个是不同的，left 是绝对的左边，而 start 会根据不同的国家习惯而改变。例如，以从右向左顺序阅读的国家，start 代表的就是右边。

实例一：第一个子控件设置水平垂直。

```
<?xml version="1.0" encoding="utf-8"?>
<LinearLayoutxmlns:android="http://schemas.android.com/apk/res/android"
xmlns:tools="http://schemas.android.com/tools"
android:layout_width="match_parent"
android:layout_height="match_parent"
android:orientation="vertical"
tools:context="com.example.icephone_1.layouttest.MainActivity">

<TextView
android:id="@+id/tx_one"
android:textSize="30sp"
android:layout_gravity="center_horizontal"   //子控件设置水平垂直
android:layout_width="wrap_content"
android:layout_height="wrap_content"
android:text="Hello World!" />

<TextView
android:id="@+id/tx_two"
android:textSize="30sp"
android:layout_width="wrap_content"
android:layout_height="wrap_content"
android:text="Hello World!" />
```

效果如图 4-1 所示。

图 4-1

实例二：设置 LinearLayout 水平垂直。

```xml
<?xml version="1.0" encoding="utf-8"?>
<LinearLayout xmlns:android="http://schemas.android.com/apk/res/android"
    xmlns:tools="http://schemas.android.com/tools"
    android:layout_width="match_parent"
    android:layout_height="match_parent"
    android:orientation="vertical"
    android:gravity="center_horizontal"   //设置 LinearLayout 水平垂直
tools:context="com.example.icephone_1.layouttest.MainActivity">

<TextView
        android:id="@+id/tx_one"
        android:textSize="30sp"
        android:layout_width="wrap_content"
        android:layout_height="wrap_content"
        android:text="Hello World!" />

<TextView
        android:id="@+id/tx_two"
        android:textSize="30sp"
        android:layout_width="wrap_content"
        android:layout_height="wrap_content"
        android:text="Hello World!" />
</LinearLayout>
```

效果如图 4-2 所示。

图 4-2

子控件大小设置如下。
属性：
- layout_height
- layout_width
- layout_weight

属性值：
- layout_height= "wrap_content" 根据子控件内容的大小决定大小
- layout_height= "match_parent" 子控件填满父容器
- layout_height= "xdp" 直接设置大小

比较特殊的用法：
- layout_height= "0dp"
- layout_weight= "1"

注意，当大小为 0dp 时，需要配合 weight 使用，表示比例。

实例：

```
<?xml version="1.0" encoding="utf-8"?>
<LinearLayoutxmlns:android="http://schemas.android.com/apk/res/android"
xmlns:tools="http://schemas.android.com/tools"
android:layout_width="match_parent"
android:layout_height="match_parent"
android:orientation="horizontal"
android:gravity="center"
tools:context="com.example.icephone_1.layouttest.MainActivity">
<TextView
android:id="@+id/tx_one"
android:textSize="30sp"
android:layout_width="0dp"
android:layout_height="wrap_content"
        android:layout_weight="1"          //设置占比例为 1
        android:text="Hello World!"
        android:background="#9c9292"/>
<TextView
        android:id="@+id/tx_two"
        android:textSize="30sp"
        android:layout_width="0dp"
        android:layout_height="wrap_content"
        android:layout_weight="1"          //设置占比例为 1
        android:text="Hello World!"
        android:background="#0d6074"/>
</LinearLayout>
```

上面的代码设置两个 TextView 的 weight 值均为 1，则这两个 TextView 各占一半空间。效果如图 4-3 所示。

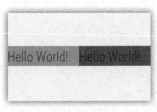

图 4-3

如果第一个设置成 2，则两个空间水平位置占比就变成了 2∶1，效果如图 4-4 所示。

图 4-4

4.2.2　RelativeLayout 相对布局详解

RelativeLayout 与 LinearLayout 严格的线性排列不同，其更随意，可以让子控件出现在整个布局的任何位置。RelativeLayout 的属性较多，但很有规律，理解了其中一个即可理解所有。

RelativeLayout 的 XML 属性、相关方法及说明，如表 4-3 所示。

表 4-3

XML 属性	相关方法	说明
android:gravity	setGravity(int)	设置布局管理器内组件的对齐方式
android:ignoreGravity	setIgnoreGravity(int)	设置哪个组件不受 gravity 属性的影响

为了控制该布局容器的各子组件的布局分布，RelativeLayout 提供了一个内部类：RelativeLayout.LayoutParams。

RelativeLayout.LayoutParams 中只能设为 boolean 的 XML 属性及说明，如表 4-4 所示。

表 4-4

XML 属性	说明
android:layout_centerHorizontal	设置该子组件是否位于布局容器的水平居中
android:layout_centerVertical	设置该子组件是否位于布局容器的纵向居中
android:layout_centerParent	设置该子组件是否位于布局容器的横纵向居中
android:layout_alignParentBottom	设置该子组件是否位于布局容器的下边框对齐
android:layout_alignParentLeft	设置该子组件是否位于布局容器的左边框对齐
android:layout_alignParentRight	设置该子组件是否位于布局容器的右边框对齐
android:layout_alignParentTop	设置该子组件是否位于布局容器的上边框对齐

RelativeLayout.LayoutParams 中属性值为其他 UI 组件 ID 的 XML 属性及说明，如表 4-5 所示。

表 4-5

XML 属性	说明
android:layout_toRightOf	控制该子组件位于给出 ID 组件的右侧
android:layout_toLeftOf	控制该子组件位于给出 ID 组件的左侧
android:layout_above	控制该子组件位于给出 ID 组件的上侧
android:layout_below	控制该子组件位于给出 ID 组件的下侧
android:layout_alignTop	控制该子组件位于给出 ID 组件的上边框对齐
android:layout_alignBottom	控制该子组件位于给出 ID 组件的下边框对齐
android:layout_alignRight	控制该子组件位于给出 ID 组件的右边框对齐
android:layout_alignLeft	控制该子组件位于给出 ID 组件的左边框对齐

RelativeLayout 常用属性值有如下几种(属性值为 true 或 false)。

- android:layout_centerHrizontal 水平居中
- android:layout_centerVertical 垂直居中
- android:layout_centerInparent 相对于父元素完全居中
- android:layout_alignParentBottom 贴紧父元素的下边缘
- android:layout_alignParentLeft 贴紧父元素的左边缘
- android:layout_alignParentRight 贴紧父元素的右边缘
- android:layout_alignParentTop 贴紧父元素的上边缘

看命名就能看出该属性的意思，例如，align 是排列，alignParent 是排列在父容器的某个位置。

实例：把 3 个 TextView 进行上中下排列，代码如下。

```xml
<?xml version="1.0" encoding="utf-8"?>
<RelativeLayoutxmlns:android="http://schemas.android.com/apk/res/android"
xmlns:tools="http://schemas.android.com/tools"
android:layout_width="match_parent"
android:layout_height="match_parent"
tools:context="com.example.icephone_1.layouttest.MainActivity">

<TextView
android:id="@+id/tx_one"
android:textSize="30sp"
android:layout_width="wrap_content"
android:layout_height="wrap_content"
android:layout_alignParentStart="true"
android:layout_alignParentLeft="true"
android:text="Hello World!"
android:background="#9c9292" />

<TextView
android:id="@+id/tx_two"
```

android:textSize="30sp"
android:layout_width="wrap_content"
android:layout_height="wrap_content"
android:layout_alignParentBottom="true"
android:layout_alignParentEnd="true"
android:layout_alignParentRight="true"
android:text="Hello World!"
android:background="#0d6074" />

<TextView
android:id="@+id/tx_three"
android:textSize="30sp"
android:layout_width="wrap_content"
android:layout_height="wrap_content"
android:layout_centerInParent="true"
android:text="Hello World!"
android:background="#a73956" />

</RelativeLayout>

效果如图 4-5 所示。

图 4-5

属性值必须为 id 的引用名["@id/id-name"]的属性有如下几种。
- android:layout_below 在某元素的下方
- android:layout_above 在某元素的上方
- android:layout_toLeftOf 在某元素的左边
- android:layout_toRightOf 在某元素的右边
- android:layout_alignTop 本元素的上边缘和某元素的上边缘对齐
- android:layout_alignLeft 本元素的左边缘和某元素的左边缘对齐
- android:layout_alignBottom 本元素的下边缘和某元素的下边缘对齐
- android:layout_alignRight 本元素的右边缘和某元素的右边缘对齐

根据另一个控件的位置来确定控件的位置。

实例：把 3 个控件排成阶梯状，代码如下。

```xml
<?xml version="1.0" encoding="utf-8"?>
<RelativeLayoutxmlns:android="http://schemas.android.com/apk/res/android"
xmlns:tools="http://schemas.android.com/tools"
android:layout_width="match_parent"
android:layout_height="match_parent"
tools:context="com.example.icephone_1.layouttest.MainActivity">

<TextView
android:id="@+id/tx_one"
android:textSize="30sp"
android:layout_width="250dp"
android:layout_height="wrap_content"
android:layout_alignStart="@+id/tx_three"
android:layout_alignLeft="@+id/tx_three"
android:layout_above="@+id/tx_three"
android:text="Hello World!"
android:background="#9c9292" />

<TextView
android:id="@+id/tx_two"
android:textSize="30sp"
android:layout_width="250dp"
android:layout_height="wrap_content"
android:layout_below="@+id/tx_three"
android:layout_alignEnd="@+id/tx_three"
android:layout_alignRight="@+id/tx_three"

android:text="Hello World!"
android:background="#0d6074" />
<TextView
android:id="@+id/tx_three"
android:textSize="30sp"
android:layout_width="wrap_content"
android:layout_height="wrap_content"

android:layout_centerInParent="true"
android:text="Hello World!"
android:background="#a73956" />

</RelativeLayout>
```

效果如图 4-6 所示。

图 4-6

属性值为具体的像素值(如 30dip、40px)的属性有如下几种。
- android:layout_marginBottom 离某元素底边缘的距离
- android:layout_marginLeft 离某元素左边缘的距离
- android:layout_marginRight 离某元素右边缘的距离
- android:layout_marginTop 离某元素上边缘的距离

需要注意的是，padding 属性与 margin 属性非常相似，这两个属性很容易混淆。

margin 属性是边缘(外边距)，指该控件距离父控件或其他控件的边距；padding 属性是填充(内边距)，指该控件内部内容，如文本或图片距离该控件的边距。效果如图 4-7 所示。

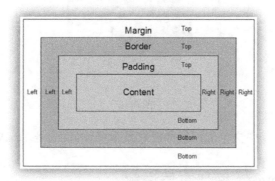

图 4-7

再例如：还是上面的代码，把两个上下控件改成相同的，效果如图 4-8 所示。

图 4-8

再给下面的控件添加一些属性，让内边距增加。
- android:paddingTop="8dp"
- android:paddingLeft="20dp"
- android:paddingStart="60dp"

完成后效果如图 4-9 所示。

图 4-9

添加 padding 属性，可以看到 TextView 中的文字位置变化了。

如果添加如下的 margin 属性呢？

- android:layout_marginTop="8dp"
- android:layout_marginLeft="20dp"
- android:layout_marginStart="20dp"

值与 padding 一样，完成后，效果如图 4-10 所示。

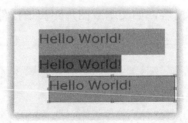

图 4-10

可以看到 TextView 中文字的位置并没有发生变化，反而是 TextView 本身发生了变化。

4.2.3 FrameLayout 框架布局详解

FrameLayout 直接继承自 ViewGroup 组件。框架布局为每个加入其中的组件创建一个空白的区域(称为一帧)，每个子组件占据一帧，这些帧会根据 gravity 属性执行自动对齐。
FrameLayout 的常用 XML 属性、相关方法及说明，如表 4-6 所示。

表 4-6

XML 属性	相关方法	说明
android:foreground	setForeground(Drawable)	设置该框架布局容器的前景图像
android:foregroundGravity	setForeGroundGraity(int)	定义绘制前景图像的 gravity 属性

实例：

```
<?xml version="1.0" encoding="utf-8"?>
<FrameLayout xmlns:android="http://schemas.android.com/apk/res/android"
    xmlns:tools="http://schemas.android.com/tools"
```

```
    android:layout_width="match_parent"
        android:layout_height="match_parent"
        tools:context="com.example.icephone_1.layouttest.MainActivity">

    <TextView
            android:id="@+id/tx_one"
            android:textSize="30sp"
            android:layout_width="300dp"
            android:layout_height="300dp"
            android:text="Hello World!"
            android:background="#9c9292" />

    <TextView
            android:id="@+id/tx_two"
            android:textSize="30sp"
            android:layout_width="200dp"
            android:layout_height="200dp"
            android:text="Hello World!"
            android:background="#1c7649" />

    <TextView
            android:id="@+id/tx_three"
            android:textSize="30sp"
            android:layout_width="100dp"
            android:layout_height="100dp"
            android:text="Hello World!"
            android:background="#a73956" />
</FrameLayout>
```

效果如图 4-11 所示。

图 4-11

4.2.4　AbsoluteLayout 绝对布局详解

AbsoluteLayout 绝对布局即 Android 不提供任何布局控制，由开发人员通过 x 坐标、y 坐标来控制组件的位置。每个组件都可指定以下两个 XML 属性：layout_x；layout_y。绝对布局已经过时,建议不使用或较少使用。

实例:

```xml
<?xml version="1.0" encoding="utf-8"?>
<AbsoluteLayout xmlns:android="http://schemas.android.com/apk/res/android"
    xmlns:tools="http://schemas.android.com/tools"
    android:layout_width="match_parent"
    android:layout_height="match_parent">
<Button
        android:id="@+id/button1"
        android:layout_width="wrap_content"
        android:layout_height="wrap_content"
        android:text="Button1"
        android:layout_y="50dp"
        android:layout_x="50dp"/>
<Button
        android:id="@+id/button2"
android:layout_width="wrap_content"
        android:layout_height="wrap_content"
android:text="Button2"
        android:layout_x="200dp"
        android:layout_y="200dp"/>
</AbsoluteLayout>
```

效果如图 4-12 所示。

图 4-12

4.2.5　TableLayout 表格布局详解

TableLayout 继承自 Linearout,本质上仍然是线性布局管理器。表格布局采用行、列的形式来管理 UI 组件,并不需要明确地声明包含多少行、多少列,而是通过添加 TableRow 及其他组件来控制表格的行数和列数。

- 每向 TableLayout 中添加一个 TableRow 就代表一行。
- 每向 TableRow 中添加一个子组件就表示一列。
- 如果直接向 TableLayout 添加组件,那么该组件将直接占用一行。

在表格布局中，可以为单元格设置如下 3 种行为方式。
- Shrinkable：该列的所有单元格的宽度可以被收缩，以保证该表格能适应父容器的宽度。
- Strentchable：该列的所有单元格的宽度可以被拉伸，以保证组件能完全填满表格空余空间。
- Collapsed：如果该列被设置为 Collapsed，那么该列的所有单元格会被隐藏。

TableLayout 的常用 XML 属性、相关方法及说明，如表 4-7 所示。

表 4-7

XML 属性	相关方法	说明
android:collapseColumns	setColumns(int, boolean)	设置需要被隐藏的列的序号，多个序号间用逗号分隔
android:shrinkColumns	setShrinkAllColumns(boolean)	设置需要被收缩的列的序号
android:stretchColumns	setStretchAllColumns(boolean)	设置允许被拉伸的列的序号

实例：

```xml
<?xml version="1.0" encoding="utf-8"?>
<TableLayout xmlns:android="http://schemas.android.com/apk/res/android"
    android:layout_width="match_parent"
    android:layout_height="match_parent">
<TableRow
        android:layout_width="match_parent"
        android:layout_height="match_parent"
        android:id="@+id/tableRow1"
>
<Button
        android:id="@+id/button1"
        android:layout_width="wrap_content"
        android:layout_height="wrap_content"
        android:text="Button1"
        android:layout_column="0"/>
<Button
        android:id="@+id/button2"
        android:layout_width="wrap_content"
        android:layout_height="wrap_content"
        android:text="Button2"
        android:layout_column="1"/>
</TableRow>

<TableRow
        android:layout_width="match_parent"
        android:layout_height="match_parent"
        android:id="@+id/tableRow2">
<Button
```

```xml
            android:id="@+id/button3"
            android:layout_width="wrap_content"
            android:layout_height="wrap_content"
            android:text="Button3"
            android:layout_column="1"
            />
    <Button
            android:id="@+id/button4"
            android:layout_width="wrap_content"
            android:layout_height="wrap_content"
            android:text="Button4"
        android:layout_column="2"/>
</TableRow>

<TableRow
        android:layout_width="match_parent"
        android:layout_height="match_parent"
        android:id="@+id/tableRow3">
    <Button
            android:id="@+id/button5"
            android:layout_width="wrap_content"
            android:layout_height="wrap_content"
            android:text="Button5"
            android:layout_column="2"/>
</TableRow>
</TableLayout>
```

效果如图 4-13 所示。

图 4-13

4.2.6 GridLayout 网格布局详解

GridLayout 是 Android 4.0 增加的网格布局控件，与之前的 TableLayout 有些相似，它

把整个容器划分为 rows * columns 个网格,每个网格可以放置一个组件。GridLayout 性能及功能都要比 TableLayout 好,例如,GridLayout 布局中的单元格可以跨越多行,而 TableLayout 则不行,此外,其渲染速度也比 TableLayout 要快。

GridLayout 提供了 setRowCount(int) 和 setColumnCount(int) 方法来控制该网格的行和列的数量。

GridLayout 常用的 XML 属性、相关方法及说明,如表 4-8 所示。

表 4-8

XML 属性	相关方法	说明
android:alignmentMode	setAlignmentMode(int)	设置该布局管理器采用的对齐模式
android:columnCount	setColumnCount(int)	设置该网格的列数量
android:columnOrderPreserved	setColumnOrderPreserved(boolean)	设置该网格容器是否保留序列号
android:roeCount	setRowCount(int)	设置该网格的行数量
android:rowOrderPreserved	setRowOrderPreserved(boolean)	设置该网格容器是否保留行序号
android:useDefaultMargins	setUseDefaultMargins(boolean)	设置该布局管理器是否使用默认的页边距

为了控制 GridLayout 布局容器中各子组件的布局分布,GridLayout 提供了一个内部类——GridLayout.LayoutParams,来控制 GridLayout 布局容器中子组件的布局分布,其 XML 属性及说明如表 4-9 所示。

表 4-9

XML 属性	说明
android:layout_column	设置该子组件在 GridLayout 的第几列
android:layout_columnSpan	设置该子组件在 GridLayout 横向上跨几列
android:layout_gravity	设置该子组件采用何种方式占据该网格的空间
android:layout_row	设置该子组件在 GridLayout 的第几行
android:layout_rowSpan	设置该子组件在 GridLayout 纵向上跨几行

实例:

```
<?xml version="1.0" encoding="utf-8"?>
<GridLayout xmlns:android="http://schemas.android.com/apk/res/android"
    android:layout_width="wrap_content"
    android:layout_height="wrap_content"
    android:layout_gravity="center"
    android:columnCount="4"
    android:orientation="horizontal">
```

```xml
<Button
        android:layout_column="3"
    android:text="/" />
<Button
        android:text="1" />
<Button
        android:text="2" />
<Button
        android:text="3" />
<Button
 android:text="*" />
<Button
        android:text="4" />
<Button
        android:text="5" />
<Button
        android:text="6" />
<Button
        android:text="-" />
<Button
        android:text="7" />
<Button
        android:text="8" />
<Button
        android:text="9" />
<Button
        android:text="+"
        android:layout_gravity="fill"
        android:layout_rowSpan="3"
        />
<Button
        android:text="0"
        android:layout_gravity="fill"
        android:layout_columnSpan="2"
        />
<Button
        android:text="00" />
<Button
        android:text="="
        android:layout_gravity="fill"
android:layout_columnSpan="3"/>
</GridLayout>
```

效果如图 4-14 所示。

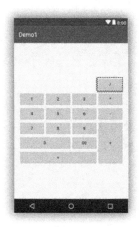

图 4-14

【单元小结】

- Android UI 布局简介。
- Android UI 常用的布局组件。

【单元自测】

1. Android 中常用的布局有哪些?
2. 下列是 AbsoluteLayout 中特有的属性的是(　　)。
   A. android:layout_height   B. android:layout_x
   C. android:layout_above   D. android:layout_toRightOf
3. 以下关于界面布局描述错误的是(　　)。
   A. RelativeLayout 中的子元素可以使用 android:layout_weight="属性"
   B. LinearLayout 中的子元素可以使用 android:layout_weight="属性"
   C. 在 TableLayout 的 TabRow 中可以添加其他控件
   D. RelativeLayout 用来表示相对布局
4. Android 中关于 View 继承关系，下面论述错误的是(　　)。
   A. ViewGroup 继承自 View
   B. AdapterView 继承自 View
   C. TableLayout 继承自 RelativeLayout
   D. LinearLayout 继承自 ViewGroup
5. 下面关于 UI 布局描述错误的是(　　)。
   A. LinearLayout 是按照横或竖的线性排列布局
   B. RelativeLayout 是按照相对位置来排列布局
   C. FrameLayout 是一块在屏幕上提前预订好的空白区域，可以填充 View 元素
   D. AbsoluteLayout 是以表格的形式布局

【上机实战】

上机目标

- 理解各种布局的排列方式及用法。
- 熟练操作布局中的属性制作项目页面。

上机练习

【问题描述】

使用线性布局、相对布局完成登录页面，如图 4-15 所示。

【问题分析】

- 如何设计父布局的方向和大小。
- 如何填充子元素在父布局中的位置。

【参考步骤】

创建 Android 工程。请参考书中前述步骤。
写 XML 布局请参考六大布局属性及介绍。

图 4-15

【拓展作业】

使用线性布局、相对布局、表格布局、网格布局完成如图 4-16 所示的计算器界面。

图 4-16

单元五　Handler 消息传递机制

课程目标

- Handler 消息传递机制介绍
- Handler 的使用

使用 Android 高级技术开发 APP

 **简 介**

Android 是不允许在子线程中进行 UI 操作的。但有些时候,我们必须在子线程中执行一些耗时任务,然后根据任务的执行结果来更新相应的 UI 控件,这该如何是好呢?针对这种情况,Android 提供了一套异步消息处理机制,完美地解决了在子线程中进行 UI 操作的问题。

5.1 Handler 消息传递机制介绍

一个 Android 应用程序被创建的时候都会创建一个 UI 主线程,但有时会有一些比较耗时的操作,为了防止阻塞 UI 主线程,我们会将耗时的操作放到子线程中进行处理,处理完之后操作 UI,但是 Android 不允许子线程操作 UI,因为 Android 的 UI 线程也是不安全的。也就是说,如果想要更新应用程序中的 UI 元素,必须在主线程中进行,否则就会出现异常。眼见为实,让我们通过一个具体的例子来验证一下吧!新建一个 HandlerTest 项目,修改 activity_main.xml 中的代码,如下所示。

```
<RelativeLayout xmlns:android="http://schemas.android.com/apk/res/android"
    xmlns:tools="http://schemas.android.com/tools"
    android:layout_width="match_parent"
    android:layout_height="match_parent"
    android:paddingBottom="@dimen/activity_vertical_margin">
<RelativeLayout
    xmlns:android="http://schemas.android.com/apk/res/android"
    android:layout_width="match_parent"
    android:layout_height="match_parent">
    <Button
        android:id="@+id/change_text"
        android:layout_width="match_parent"
        android:layout_height="wrap_content"
        android:text="Change    Text"/>
    <TextView
        android:id="@+id/text"
        android:layout_width="wrap_content"
        android:layout_height="wrap_content"
        android:layout_centerInParent="true"
        android:text="Hello world"
        android:textSize="20sp" />
</RelativeLayout>
```

布局文件中定义了两个控件,TextView 用于显示 Hello world 字符串,Button 用于改变 TextView 中显示的内容,我们希望在单击 Button 按钮后可以把 TextView 中显示的字符串改成 Hello Android。接下来修改 MainActivity 中的代码,如下所示。

```
public class MainActivity extends AppCompatActivity implements View.OnClickListener {
    private TextView text;
    @Override
    protected void onCreate(Bundle savedInstanceState) {
        super.onCreate(savedInstanceState);
        setContentView(R.layout.activity_main);
        text    =    (TextView)findViewById(R.id.text);
        Button changeText =    (Button)    findViewById(R.id.change_text);
        changeText.setOnClickListener(this);
    }
    public void onClick(View v) {
        switch (v.getId()) {
            case R.id.change_text:
                new Thread(new Runnable() {
                    @Override
                    public void run() {
                        text.setText("Hello Android");
                    }
                }).start();
                break;
            default:
                break;
        }
    }
}
```

我们在 ChangeText 按钮的单击事件中开启了一个子线程，然后在子线程中调用 TextView 的 setText()方法将显示的字符串改成 Hello Android。我们是在子线程中更新 UI 的。现在运行程序并单击 ChangeText 按钮，发现程序崩溃了，如图 5-1 所示。

图 5-1

5.2 Handler 的使用

对于上述这种情况，Android 提供了一套异步消息处理机制，完美地解决了在子线程中进行 UI 操作的问题。修改 MainActivity 中的代码，如下所示。

```
public class MainActivity extends AppCompatActivity implements View.OnClickListener {
    public static final int UPDATE_TEXT = 1;
    private TextView text;
    private Handler Handler;
    private Button changeText;
    @Override
    protected void onCreate(Bundle savedInstanceState) {
        super.onCreate(savedInstanceState);
        setContentView(R.layout.activity_main);
        text = (TextView) findViewById(R.id.text);
        changeText=findViewById(R.id.change_text);
```

```
            Handler = new Handler() {
                public void HandlerMessage(Message msg) {
                    switch (msg.what) {
                        case UPDATE_TEXT:
                            // 在这里可以进行 UI 操作
                            text.setText("Hello Android");
                            break;
                        default:
                            break;
                    }
                }
            };
            changeText.setOnClickListener(this);
    }
    public void onClick(View v) {
        switch (v.getId()) {
            case R.id.change_text:
                new Thread(new Runnable() {
                    @Override
                    public void run() {
                        Message message = new Message();
                        message.what = UPDATE_TEXT;
                        Handler.sendMessage(message); // 将 Message 对象发送出去
                    }
                }).start();
                break;
            default:
                break;
        }
    }
}
```

这里我们先定义了一个整型常量 UPDATE_TEXT，用于表示更新 TextView 这个动作，然后新增一个 Handler 对象，并重写父类的 HandlerMessage()方法，在这里对具体的 Message 进行处理。如果发现 Message 的 what 字段的值等于 UPDATE_TEXT，则将 TextView 显示的内容改成 Hello Android。这次我们并没有在子线程中直接进行 UI 操作，而是创建了一个 Message 对象，并将 what 字段的值指定为 UPDATE_TEXT，然后调用 Handler 的 sendMessage()方法将这条 Message 发送出去。很快，Handler 就会收到这条 Message，并在 HandlerMessage()方法中对它进行处理。现在重新运行程序，可以看到显示的是 Hello World。然后单击 Change Text 按钮，显示的内容就被替换成 Hello Android，如图 5-2 所示。

图 5-2

Android 中的异步消息处理有 Message、Handler、MessageQueue 和 Looper。其中，Message 和 Handler 我们已经接触过了，而 MessageQueue 和 Looper 还是全新的概念，下面就对这 4 个部分进行简要的介绍。

- Message 是在线程之间传递的消息，它可以在内部携带少量的信息，用于在不同线程之间交换数据。
- Handler 顾名思义是处理者的意思，它主要用于发送和处理消息。发送消息一般是使用 Handler 的 sendMessage()方法，而发出的消息经过一系列的辗转处理后，最终会传递到 Handler 的 HandlerMessage()方法中。
- MessageQueue 是消息队列的意思，它主要用于存放所有通过 Handler 发送的消息。这部分消息会一直存在于消息队列中等待被处理。每个线程中只有一个 MessageQueue 对象。
- Looper 是每个线程中的 MessageQueue 的管家，调用 loop()方法后，就会进入一个无限循环当中，然后每当发现 MessageQueue 中存在一条消息时，就会将它取出，并传递到 Handler 的 HandlerMessage()方法中。每个线程中也只有一个 Looper 对象。

总结

我们再来把异步消息处理的整个流程梳理一遍。首先需要在主线程中创建一个 Handler 对象，并重写 HandlerMessage()方法。然后当子线程中需要进行 UI 操作时就创建一个 Message 对象，并通过 Handler 将这条消息发送出去。最后这条消息会被添加到 MessageQueue 的队列中等待被处理，而 Looper 则会一直尝试从 MessageQueue 中取出后分发回 Handler 的 HandlerMessage()方法中。由于 Handler 是在主线程中创建的，所以此时 HandlerMessage()方法中的代码也会在主线程中运行，于是我们在这里就可以安心地进行 UI 操作了。

【单元小结】

- Handler 消息的传递机制。
- Message 的使用。

【单元自测】

1. 什么是 Handler？
2. 什么是消息机制？
3. 为什么不能在子线程中访问 UI？

【上机实战】

上机目标

- 理解如何使用 Handler。
- 理解如何使用 Message。

上机练习

【问题描述】
如何使用 Handler 在子线程中进行 UI 操作?

【问题分析】
由于 Android 在子线程中无法进行 UI 操作,而有时候我们又有需求。

【参考步骤】
- 创建一个 Handler,重写 HandlerMessage()方法,根据 msg.what 信息判断,接收对应的信息,再在这里更新 UI。
- 创建一个 Message 对象,设置 what 标志及数据。
- 通过 sendMessage 进行投递消息。

```
private Handler handler = new Handler(){      //创建一个 Message 对象
    @Override
    public void HandlerMessage(Message msg) {
        super.HandlerMessage(msg);
        switch (msg.what) {                   //判断标志位
            case 1:
                /**
                  获取数据,更新 UI
                */
                break;
        }
    }
};}
Message msg =new Message();                   //创建一个 Message 对象
msg.what=1;                                   //设置 what 标志及数据
handler.sendMessage(msg);                     //通过 sendMessage 进行投递消息
```

【拓展作业】

1. 使用 Handler 在子线程中通过与 UI 通信来更新 UI 界面。
2. 思考如何使用 Handler 进行数据传递?并查询资料进行自我学习。

单元六 UI 中级控件

课程目标

- RadioGroup 和 RadioButton
- CheckBox
- Dialog
- Spinner
- ListView 和 BaseAdapter
- GridView

 **简 介**

前面我们已经学习了 Android 的几个基本控件,发现 Android 控件开发有一个共同的特征,就是在 XML 中定义控件的基本属性,而控件的事件及控制在 Activity 中定义。本单元我们将继续学习 Android 控件的开发与应用。

应用程序的人机交互界面由众多 Android 控件组成,前面所有示例讲解的都是最简单的控件的应用,因此在实际应用中,我们见到的控件还有很多。打开我们的手机会发现,电话本是如何实现的?增加通讯录时,下拉框是如何实现的?单选按钮又是如何实现的?

6.1 RadioButton、RadioGroup

单选按钮相信大家都不陌生吧。Android 平台也提供了单选按钮的组件,在 Android 中也可以通过 RadioButton、RadioGroup 组合起来完成一个单选按钮效果,如图 6-1 所示。

图 6-1

用户选择一项喜欢的颜色后,单击"确定"按钮,会通过 Toast 对话框来提示我们的选择,它是如何实现的呢?下面是对单选按钮在 XML 的设置和代码中的具体实现。

Activity_main.xml:

```
<RelativeLayout xmlns:android="http://schemas.android.com/apk/res/android"
    xmlns:tools="http://schemas.android.com/tools"
    android:layout_width="match_parent"
    android:layout_height="match_parent"
    android:paddingBottom="@dimen/activity_vertical_margin"
```

```xml
        android:paddingLeft="@dimen/activity_horizontal_margin"
        android:paddingRight="@dimen/activity_horizontal_margin"
        android:paddingTop="@dimen/activity_vertical_margin"
        tools:context=".MainActivity" >
<LinearLayout
        android:layout_width="fill_parent"
        android:layout_height="fill_parent"
        android:orientation="vertical" >
<!-- 一个 RadioGroup 组中有多个 radio -->
<RadioGroup
            android:id="@+id/myGroup"
            android:layout_width="fill_parent"
            android:layout_height="wrap_content" >
<RadioButton
                android:id="@+id/b1"
                android:layout_width="wrap_content"
                android:layout_height="wrap_content"
                android:text="@string/red" />
<RadioButton
                android:id="@+id/b2"
                android:layout_width="wrap_content"
                android:layout_height="wrap_content"
                android:text="@string/yellow" />
<RadioButton
                android:id="@+id/b3"
                android:layout_width="wrap_content"
                android:layout_height="wrap_content"
                android:text="@string/blue" />
<RadioButton
                android:id="@+id/b4"
                android:layout_width="wrap_content"
                android:layout_height="wrap_content"
                android:text="@string/purple" />
<RadioButton
                android:id="@+id/b5"
                android:layout_width="wrap_content"
                android:layout_height="wrap_content"
                android:text="@string/black" />
</RadioGroup>
<Button
            android:id="@+id/btn"
            android:layout_width="100dip"
            android:layout_height="wrap_content"
            android:text="@string/yes" />
</LinearLayout>
</RelativeLayout>
```

在控制文件 MainActivity 中，代码如下。

```java
package com.example.administrator.myapplication;
import android.support.v7.app.AppCompatActivity;
import android.os.Bundle;
import android.view.Gravity;
import android.view.View;
import android.widget.Button;
import android.widget.RadioButton;
import android.widget.RadioGroup;
import android.widget.Toast;

public class MainActivity extends AppCompatActivity implements RadioGroup.OnCheckedChangeListener, View.OnClickListener {

    private RadioGroup myGroup;
    private RadioButton b1, b2, b3, b4, b5;
    private Button btn;
    private String mycheck;

    @Override
    protected void onCreate(Bundle savedInstanceState) {
        super.onCreate(savedInstanceState);
        setContentView(R.layout.activity_main);
        myGroup = (RadioGroup) this.findViewById(R.id.myGroup);
        b1 = (RadioButton) this.findViewById(R.id.b1);
        b2 = (RadioButton) this.findViewById(R.id.b2);
        b3 = (RadioButton) this.findViewById(R.id.b3);
        b4 = (RadioButton) this.findViewById(R.id.b4);
        b5 = (RadioButton) this.findViewById(R.id.b5);
        myGroup.setOnCheckedChangeListener(this);
        btn = (Button) this.findViewById(R.id.btn);
        btn.setOnClickListener(this);

    }

    @Override
    public void onCheckedChanged(RadioGroup group, int checkedId) {
        if (checkedId == b1.getId())
        {
            mycheck = b1.getText().toString();
        }
        else if (checkedId == b2.getId())
        {
            mycheck = b2.getText().toString();
        }
        else if (checkedId == b3.getId())
        {
```

```
                mycheck = b3.getText().toString();
            }
            else if (checkedId == b4.getId())
            {
                mycheck = b4.getText().toString();
            }
            else
            {
                mycheck = b5.getText().toString();
            }

        }

        @Override
        public void onClick(View v) {
            if (v.getId() == R.id.btn)
            {
                myPlay("我喜欢：" + mycheck);
            }
        }
        public void myPlay(String message)
        {
            Toast t = Toast.makeText(this, message, Toast.LENGTH_SHORT);
            t.setGravity(Gravity.TOP,0,300);
            t.show();
        }
}
```

以上实现当用户选择自己喜欢的颜色后，再单击"确定"按钮时，以 Toast 提示。其实在这一过程中，有两个动作的产生：一是，每一个单选按钮被选中后，产生一个 onCheckedChanged 事件，这个事件中，将获得当前选中的按钮的文本值，并给定到一个指定的变量中；二是，单击"确定"按钮将所选择的对象值显示。

6.2 CheckBox

上面操作实现了单项选择，如果在实际开发中要求选择我们的爱好，但一个人的爱好又不止一项，那么就用到复选框了，如图 6-2 所示。

有些时候，我们要得到所选择对象的编号以方便存入数据库，如图 6-3 所示。

 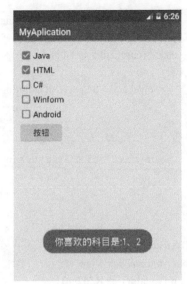

图 6-2　　　　　　　　图 6-3

具体实现的代码如下。

Activity_main.xml：

```xml
<RelativeLayout xmlns:android="http://schemas.android.com/apk/res/android"
    xmlns:tools="http://schemas.android.com/tools"
    android:layout_width="match_parent"
    android:layout_height="match_parent"
    android:paddingBottom="@dimen/activity_vertical_margin"
    android:paddingLeft="@dimen/activity_horizontal_margin"
    android:paddingRight="@dimen/activity_horizontal_margin"
    android:paddingTop="@dimen/activity_vertical_margin"
    tools:context=".MainActivity" >
<LinearLayout
        android:layout_width="fill_parent"
        android:layout_height="fill_parent"
        android:orientation="vertical" >
<CheckBox
        android:id="@+id/ck1"
        android:layout_width="wrap_content"
        android:layout_height="wrap_content"
        android:text="@string/ck1" />
<CheckBox
        android:id="@+id/ck2"
        android:layout_width="wrap_content"
        android:layout_height="wrap_content"
        android:text="@string/ck2" />
<CheckBox
        android:id="@+id/ck3"
```

```xml
                android:layout_width="wrap_content"
                android:layout_height="wrap_content"
                android:text="@string/ck3" />
<CheckBox
                android:id="@+id/ck4"
                android:layout_width="wrap_content"
                android:layout_height="wrap_content"
                android:text="@string/ck4" />
<CheckBox
                android:id="@+id/ck5"
                android:layout_width="wrap_content"
                android:layout_height="wrap_content"
                android:text="@string/ck5" />
<Button
                android:id="@+id/btn"
                android:layout_width="100dip"
                android:layout_height="wrap_content"
                android:text="@string/btn" />
</LinearLayout>
</RelativeLayout>
```

MainActivity.java：

```java
package com.example.administrator.myapplication;

import android.support.v7.app.AppCompatActivity;
import android.os.Bundle;
import android.view.View;
import android.widget.Button;
import android.widget.CheckBox;
import android.widget.CompoundButton;
import android.widget.Toast;

public class MainActivity extends AppCompatActivity implements View.OnClickListener, CompoundButton.OnCheckedChangeListener {
    private CheckBox ck1, ck2, ck3, ck4, ck5;
    private Button btn;
    private int c1 = 0;
    private int c2 = 0;
    private int c3 = 0;
    private int c4 = 0;
    private int c5 = 0;
    @Override
    protected void onCreate(Bundle savedInstanceState) {
        super.onCreate(savedInstanceState);
        setContentView(R.layout.activity_main);
        ck1 = (CheckBox) this.findViewById(R.id.ck1);
```

```java
            ck2 = (CheckBox) this.findViewById(R.id.ck2);
            ck3 = (CheckBox) this.findViewById(R.id.ck3);
            ck4 = (CheckBox) this.findViewById(R.id.ck4);
            ck5 = (CheckBox) this.findViewById(R.id.ck5);
            btn = (Button) this.findViewById(R.id.btn);
            btn.setOnClickListener(this);
            ck1.setOnCheckedChangeListener(this);
            ck2.setOnCheckedChangeListener(this);
            ck3.setOnCheckedChangeListener(this);
            ck4.setOnCheckedChangeListener(this);
            ck5.setOnCheckedChangeListener(this);
    }
    // 按钮上的事件
        @Override
        public void onClick(View v) {
            String message = new String();
            if (c1 != 0)
            {
                //message = message + "、" + ck1.getText().toString();
                message = message + "、" + c1;
            }
            if (c2 != 0)
            {
                //message = message + "、" +   ck2.getText().toString();
                message = message + "、" + c2;
            }
            if (c3 != 0)
            {
                //message = message + "、" +   ck3.getText().toString();
                message = message + "、" + c3;
            }
            if (c4 != 0)
            {
                //message = message + "、" +   ck4.getText().toString();
                message = message + "、" + c4;
            }
            if (c5 != 0)
            {
                //message = message + "、" +   ck5.getText().toString();
                message = message + "、" + c5;
            }
            String newMessage = message.substring(1, message.length());
            myPlay("你喜欢的科目是:" + newMessage);        }
    //多选框选中与没有选中事件
        @Override
        public void onCheckedChanged(CompoundButton buttonView, boolean isChecked) {
            if (ck1.isChecked())
```

```
            {
                c1 = 1;
            }
            if (!ck1.isChecked())
            {
                c1 = 0;
            }
            if (ck2.isChecked())
            {
                c2 = 2;
            }
            if (!ck2.isChecked())
            {
                c2 = 0;
            }
            if (ck3.isChecked())
            {
                c3 = 3;
            }
            if (!ck3.isChecked())
            {
                c3 = 0;
            }
            if (ck4.isChecked())
            {
                c4 = 4;
            }
            if (!ck4.isChecked())
            {
                c4 = 0;
            }
            if (ck5.isChecked())
            {
                c5 = 5;
            }
            if (!ck5.isChecked())
            {
                c5 = 0;
            }
        }
        //Toast 提示方法
        public void myPlay(String message)
        {
            Toast t = Toast.makeText(this, message, Toast.LENGTH_SHORT);
            t.show();
        }
}
```

想一想，如下代码为什么要这样写呢？

```
if (ck1.isChecked())
    {
        c1 = 1;
    }
    if (!ck1.isChecked())
    {
        c1 = 0;
    }
```

如果不知道，说明你没有认真去练习 checkBox，自己去找一找答案吧！

需要注意的是，上面单选按钮实现的是 OnCheckedChangeListener 接口，而这个接口在包 android.widget.RadioGroup.OnCheckedChangeListener 中。复选框实现的也是 OnCheckedChangeListener 接口，而这个接口在包 android.widget.CompoundButton.OnChecked-ChangeListener 中。

6.3 对话框(DiaLog)

Android 应用程序中，有时在操作某些动作时，为了给用户更友好的提示，用到了弹出式对话框，即 Dialog 对话框。Android 中实现对话框可以使用 AlertDialog.Builder 类，还可以自定义对话框。图 6-4 所示是实现用户退出系统时的友好提示。

图 6-4

具体代码如下：

```
package com.example.administrator.myapplication;
import android.app.AlertDialog;
import android.app.Dialog;
```

```java
import android.content.DialogInterface;
import android.support.v7.app.AppCompatActivity;
import android.os.Bundle;
import android.view.View;
import android.widget.Button;
public class MainActivity extends AppCompatActivity {
    private Button out;
    @Override
    protected void onCreate(Bundle savedInstanceState) {
        super.onCreate(savedInstanceState);
        setContentView(R.layout.activity_main);
        initView();
    }
    //初始化控件方法
    private void initView()
    {
        out=(Button) this.findViewById(R.id.btn);
        //加事件
        out.setOnClickListener(myout);
    }
    View.OnClickListener myout=new View.OnClickListener() {
        @Override
        public void onClick(View arg0)
        {
            //定义对话框对象
Dialog dialog=new AlertDialog.Builder(MainActivity.this)
.setTitle("退出提示")   //设置对话框的标题
.setMessage("你真的要退出吗？")   //设置对话框的提示内容
.setPositiveButton("确定",new DialogInterface.OnClickListener() //确定按钮事件
                {
                    @Override
                    public void onClick(DialogInterface arg0, int arg1)
                    {
                        //这里是退出后处理的地方
                    }
                })
.setNegativeButton("取消",new DialogInterface.OnClickListener() //取消按钮事件
                {
                    @Override
                    public void onClick(DialogInterface dialog, int which)
                    {
                        //这里是取消后事件处理的地方
                    }
                })
                .create();//创建这个对象
```

```
                dialog.show(); //显示
            }
        };
    }
```

在实际应用中，往往结合线程使用，让用户体验效果更强，如图 6-5(a)和(b)所示。

(a)

(b)

图 6-5

定义一个 ProgressDialog 对象 m_dialog，具体代码如下。

```
package com.example.administrator.myapplication;
import android.app.AlertDialog;
import android.app.ProgressDialog;
import android.content.DialogInterface;
import android.content.Intent;
import android.support.v7.app.AppCompatActivity;
import android.os.Bundle;
import android.view.View;
import android.widget.Button;
public class MainActivity extends AppCompatActivity {
    private Button out;
    @Override
    protected void onCreate(Bundle savedInstanceState) {
        super.onCreate(savedInstanceState);
        setContentView(R.layout.activity_main);
        initView();
    }
    //初始化控件方法
```

```java
private void initView()
{
    out=(Button) this.findViewById(R.id.btn);
    //加事件
    out.setOnClickListener(myout);
}

private AlertDialog dialog;
private ProgressDialog m_dialog;
View.OnClickListener myout=new View.OnClickListener() {

    @Override
    public void onClick(View arg0)
    {
        //定义对话框对象
        dialog=new AlertDialog.Builder(MainActivity.this)
        .setTitle("退出提示")                    //设置对话框的标题
        .setMessage("你真的要退出吗？")          //设置对话框的提示内容
        .setPositiveButton("确定",new DialogInterface.OnClickListener()    //确定按钮事件
        {
            final CharSequence strDialogTitle = "友情提示";
            final CharSequence strDialogBody = "退出进行中……";
            @Override
            public void onClick(DialogInterface arg0, int arg1)
            {
                m_dialog=ProgressDialog.show(MainActivity.this, strDialogTitle, strDialogBody);
                new Thread(new Runnable() {
                    @Override
                    public void run() {
                        try
                        {
                            Thread.sleep(5000);
                        }
                        catch (Exception e)
                        {
                            e.printStackTrace();
                        }
                        finally
                        {
                            //这里是退出后处理的地方
                            Intent intent = new Intent();
                            //跳转到第二页面中(SecondActivity 是第二个 Activity)
                            intent.setClass(getApplicationContext(), SecondActivity.class);
                            startActivity(intent);
                            dialog.dismiss();
                        }
                    }
```

```
                            }
                        }).start();                          //这里是退出后处理的地方
                    }
                })
                .setNegativeButton("取消",new DialogInterface.OnClickListener() //取消按钮事件
                {
                        @Override
                        public void onClick(DialogInterface dialog, int which)
                        {
                            //这里是取消后事件处理的地方
                        }
                })
                .create();
        dialog.show();
    }
};
}
```

6.4 Spinner

当我们在某个网站上注册账号时,可能要提供性别、生日、城市等信息,而网站开发人员为了方便用户,在城市这个选项上采用下拉框来供用户选择,下面我们就来学习在 Android 中下拉框是如何实现的,如图 6-6(a)和(b)所示。

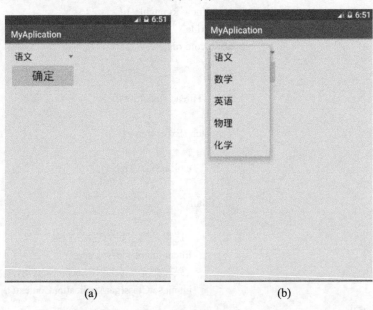

图 6-6

下拉框功能是如何实现的呢？它的具体实现如下。

Activity_main.xml：

```xml
<RelativeLayout xmlns:android="http://schemas.android.com/apk/res/android"
    xmlns:tools="http://schemas.android.com/tools"
    android:layout_width="match_parent"
    android:layout_height="match_parent"
    android:paddingBottom="@dimen/activity_vertical_margin"
    android:paddingLeft="@dimen/activity_horizontal_margin"
    android:paddingRight="@dimen/activity_horizontal_margin"
    android:paddingTop="@dimen/activity_vertical_margin"
    tools:context=".MainActivity" >
<LinearLayout
        android:layout_width="fill_parent"
        android:layout_height="fill_parent"
        android:orientation="vertical" >
<Spinner
            android:id="@+id/mysp"
            android:layout_width="150dip"
            android:layout_height="wrap_content"
            android:gravity="center_horizontal"/>
<Button
            android:id="@+id/btn"
            android:layout_width="100dip"
            android:layout_height="wrap_content"
            android:text="确定" />
</LinearLayout>
</RelativeLayout>
```

从上面布局 XML 文件我们可以看出，这只是一个下拉框的声明，并没有具体的下拉框内容。那么下拉框的内容来自哪儿呢？其实现是在控制类中，具体如下。

```java
package com.example.administrator.myapplication;
import android.support.v7.app.AppCompatActivity;
import android.os.Bundle;
import android.view.View;
import android.widget.AdapterView;
import android.widget.ArrayAdapter;
import android.widget.Button;
import android.widget.Spinner;
import android.widget.Toast;

public class MainActivity extends AppCompatActivity implements AdapterView.OnItemSelectedListener, View.OnClickListener {
    private Spinner mysp;
    private Button btn;
    private String myck;
    private String[] mymessage ={ "语文","数学","英语","物理","化学" };
```

```java
        private ArrayAdapter adapter;
        @Override
        protected void onCreate(Bundle savedInstanceState) {
            super.onCreate(savedInstanceState);
            setContentView(R.layout.activity_main);
            // 得到下拉框对象
            mysp = (Spinner) this.findViewById(R.id.mysp);
            // 把内容绑定到 ArrayAdapter 上
            adapter = new ArrayAdapter(this, android.R.layout.simple_spinner_item,
                    mymessage);
            // 设置下拉框的风格
            adapter.setDropDownViewResource(android.R.layout.simple_spinner_dropdown_item);
            // 把 ArrayAdapter 绑定到下拉框上
            mysp.setAdapter(adapter);
            btn = (Button) this.findViewById(R.id.btn);
            mysp.setOnItemSelectedListener(this);
            btn.setOnClickListener(this);
        }
        /***
         * 下拉框选择事件
         */
        @Override
        public void onItemSelected(AdapterView<?> parent, View view, int position, long id) {
            myck = mymessage[position];
            parent.setVisibility(View.VISIBLE);
        }
        @Override
        public void onNothingSelected(AdapterView<?> parent) {
        }
        /***
         * 按钮上的事件
         */
        @Override
        public void onClick(View v) {
            myPlay("你喜欢："+myck);
        }
        //Toast 事件
        public void myPlay(String message)
        {
            Toast t = Toast.makeText(this, message, Toast.LENGTH_SHORT);
            t.show();
        }
    }
```

> **注意**
>
> Spinner 下拉框实现的是 OnItemSelectedListener 接口，并且这个接口有两个方法来实现，我们在常规开发中，只需要实现操作 onItemSelected 方法即可。

6.5 ListView

6.5.1 简介

- Android 中的一种列表视图组件。
- 集合多个"项"(称为 Item)，并且以列表的形式展示。
- 继承自 AdapterView 抽象类，类图关系如图 6-7 所示。

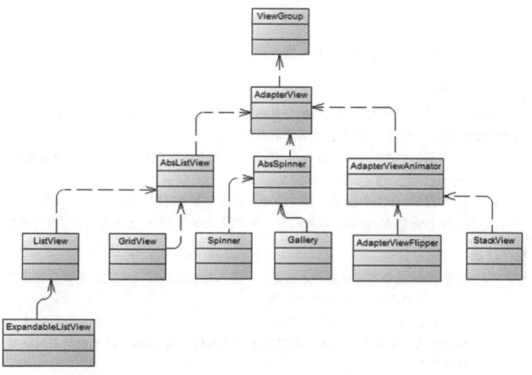

图 6-7

6.5.2 工作原理

1. 本质原理

- ListView 仅作为容器(列表)，用于装载和显示数据(即列表项 Item)，而容器内的具体数据(列表项 Item)由适配器(Adapter)提供。
- 适配器(Adapter)作为 View 和数据之间的桥梁和中介，将数据映射到要展示的 View 中。
- 当需要显示数据时，ListView 会向 Adapter 取出数据，从而加载显示，具体如图 6-8 所示。

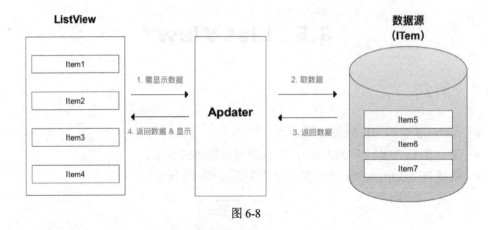

图 6-8

结论：ListView 负责以列表的形式显示 Adapter 提供的内容。

2. 缓存原理

试想一个场景：把所有数据集合的信息都加载到 ListView 上显示，若 ListView 要为每个数据都创建一个视图，那么会占用非常多的内存。

- 为了节省空间和时间，ListView 不会为每一个数据创建一个视图，而是采用了 Recycler 组件，用于回收和复用 View。
- 当屏幕需显示 x 个 Item 时，ListView 会创建 $x+1$ 个视图；当第 1 个 Item 离开屏幕时，此 Item 的 View 被回收至缓存，入屏的 Item 的 View 会优先从该缓存中获取。

 注意

- 只有 Item 完全离开屏幕后才可复用，这也是 ListView 要创建比屏幕需显示视图多一个的原因——缓冲显示视图。
- 第 1 个 Item 离开屏幕是有过程的，第 1 个 Item 的下半部分和第 8 个 Item 上半部分会同时在屏幕中显示，此时仍无法使用缓存的 View，只能继续用新创建的视图 View。

假设：若屏幕只能显示 5 个 Item，那么 ListView 只会创建(5+1)个 Item 的视图；当第 1 个 Item 完全离开屏幕后才会回收至缓存从而复用(用于显示第 7 个 Item)，如图 6-9、图 6-10、图 6-11 所示。

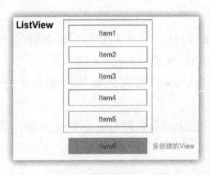

图 6-9

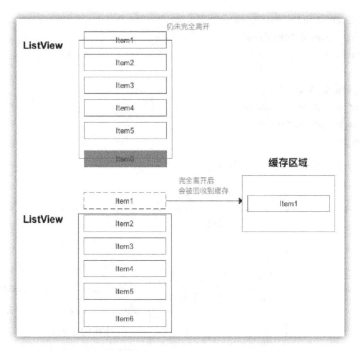

图 6-10

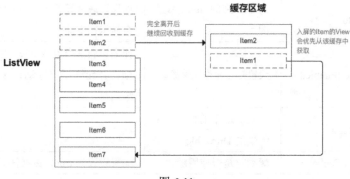

图 6-11

6.5.3 具体使用

1. 生成方式

- 直接用 ListView 进行创建。
- 让 Activity 继承 ListActivity。

2. xml 文件配置信息

代码如下。

```
<LinearLayout xmlns:android="http://schemas.android.com/apk/res/android"
    xmlns:tools="http://schemas.android.com/tools"
    android:layout_width="match_parent"
```

```
            android:layout_height="match_parent"
            android:background="#FFE1FF"
            android:orientation="vertical">
<ListView
android:id="@+id/listView1"
android:layout_width="match_parent"
android:layout_height="match_parent" />
</LinearLayout>
```

ListView 的常用属性和说明如表 6-1 所示。

表 6-1

属性	说明	备注
android:choiceMode	列表的选择行为，默认 none 没有选择行为	选择方式：①none，不显示任何选中项；②singleChoice，允许单选；③multipleChoice，允许多选；④multipleChoiceModal，允许多选(把 Activity 中 adapter 的第二个参数改成支持选择的布局)
android:drawSelectorOnTop		如果该属性设置为 true，则选中的列表项将会显示在上面
android:listSelector	为单击到的 Item 设置图片	如果该属性设置为 true，则选中的列表项将会显示在上面
android：fastScrollEnabled	设置是否允许快速滚动	如果该属性设置为 true，则将会显示滚动图标，并允许用户拖动该滚动图标进行快速滚动
android：listSelector	指定被选中的列表项上绘制的 Drawable	
android：scrollingCache	滚动时是否使用缓存	如果设置为 true，则在滚动时将会使用缓存
android：stackFromBottom	设置是否从底端开始排列列表项	
android：transcriptMode	指定列表添加新的选项的时候，是否自动滑动到底部，显示新的选项	①disabled，取消 transcriptMode 模式。②默认的 normal，当接受到数据集合改变的通知，并且仅当最后一个选项已经显示在屏幕上时，自动滑动到底部。③alwaysScroll，无论当前列表显示什么选项，列表都将会自动滑动到底部显示最新的选项

Listview 提供的 XML 属性及说明如表 6-2 所示。

表 6-2

属性	说明	备注
android:divider	设置 List 列表项的分割条(可用颜色分割，也可用图片(Drawable)分割	不设置列表之间的分割线，可设置属性为@null
android:dividerHeight	用于设置分割条的高度	
android:background	设置列表的背景	
android:entries	指定一个数组资源，Android 将根据该数组资源生成 ListView	
android:footerDividerEnabled	如果设置成 false，则不在 footer View 之前绘制分割条	
andorid:headerDividerEnabled	如果设置成 false，则不再 header View 之前绘制分割条	

6.5.4 Adapter 介绍

Adapter 本身是一个接口，Adapter 接口及其子类的继承关系如图 6-12 所示。

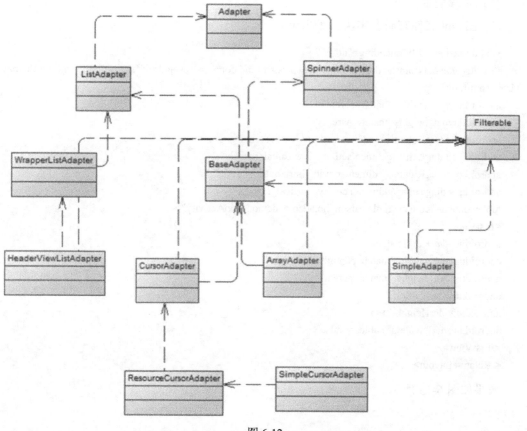

图 6-12

- Adapter 接口派生了 ListAdapter 和 SpinnerAdapter 两个子接口。其中，ListAdapter 为 AbsAdapter 提供列表项，而 SpinnerAdapter 为 AbsSpinner 提供列表项。
- ArrayAdapter、SimpleAdapter、SimpleCursorAdapter、BaseAdapter 都是常用的实现适配器的类。
- ArrayAdapter：简单、易用的 Adapter，用于将数组绑定为列表项的数据源，支持泛型操作。
- SimpleAdapter：功能强大的Adapter，用于将XML中的控件绑定为列表项的数据源。
- SimpleCursorAdapter：与 SimpleAdapter 类似，用于绑定游标(直接从数据库取出数据)作为列表项的数据源。
- BaseAdapter：可自定义 ListView，通常用于被扩展。扩展 BaseAdapter 可以对各个列表项进行最大程度的定制。

6.5.5 常用适配器介绍

1. ArrayAdapter 详解

ArrayAdapter 数组适配器用于绑定格式单一的数据，数据源可以是集合或者数组列表视图(ListView)以垂直的形式列出需要显示的列表项。

操作步骤如下。

(1) 在 xml 文件布局上实现 ListView。

```xml
<?xml version="1.0" encoding="utf-8"?>
<RelativeLayout xmlns:android="http://schemas.android.com/apk/res/android" xmlns:tools="http://schemas.android.com/tools"
    android:layout_width="match_parent"
    android:layout_height="match_parent"
    android:paddingBottom="@dimen/activity_vertical_margin"
    android:paddingLeft="@dimen/activity_horizontal_margin"
    android:paddingRight="@dimen/activity_horizontal_margin"
    android:paddingTop="@dimen/activity_vertical_margin"
    tools:context="com.example.carson_ho.adapte_demo.MainActivity">
    <ListView
    android:id="@+id/list_item"
    android:layout_width="match_parent"
    android:layout_height="match_parent"
    android:divider="#f00"
    android:dividerHeight="1sp"
    android:headerDividersEnabled="false">
    </ListView>
</RelativeLayout>
```

效果如图 6-13 所示。

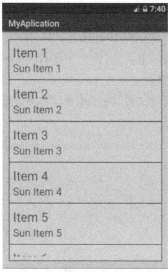

图 6-13

(2) 在 MainActivity 中定义一个链表，将所要展示的数据存放在里面。
(3) 构造 ArrayAdapter 对象，设置适配器。
(4) 将 LsitView 绑定到 ArrayAdapter 上。

```
publicclassMainActivityextendsAppCompatActivity{


@Override
protectedvoidonCreate(Bundle savedInstanceState){
super.onCreate(savedInstanceState);
    setContentView(R.layout.activity_main);

        ListView listView = (ListView) findViewById(R.id.list_item);
//定义一个链表用于存放要显示的数据
final List<String> adapterData = new ArrayList<String>();
//存放要显示的数据
for (int i = 0; i <20; i++) {
            adapterData.add("ListItem" + i);
        }
//创建 ArrayAdapter 对象 adapter 并设置适配器
        ArrayAdapter<String> adapter = new ArrayAdapter<String>(this,
                android.R.layout.simple_list_item_1, adapterData);
//将 LsitView 绑定到 ArrayAdapter 上
        listView.setAdapter(adapter);
    }
}
```

创建 ArrayAdapter 对象要指定以下 3 个参数。
- context：代表方位 Android 应用的接口。
- textViewRseourceId：资源 ID，代表一个 TextView。
- 数组：列表项展示的数据。

(5) 在 xml 文件布局添加资源文件 TextView，该 TextView 组件将作为列表项的组件。

```
<?xml version="1.0" encoding="utf-8"?>
<TextViewxmlns:android="http://schemas.android.com/apk/res/android"
android:layout_width="match_parent"
android:layout_height="wrap_content">
android:textSize="24sp"
</TextView>
```

最终效果如图 6-14 所示。

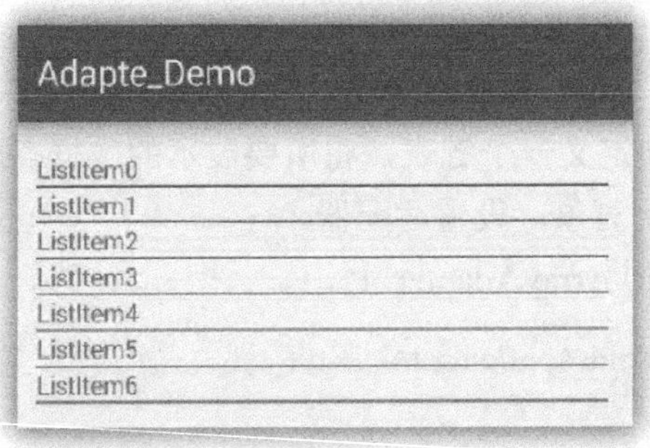

图 6-14

ArrayAdapter 较为简单、易用，但每个列表项只能是 TextView，功能实现的局限性非常大。

2. SimpleAdapter 详解

SimpleAdapter 是扩展性非常好的一种适配器，相比较 ArrayAdapter 而言，SimleAdapter 不仅可以显示文本信息，还可以显示更多的内容，如图片、按钮等，可以说我们在日常开发中使用 SimpleAdapter 的频率是比较高的。

操作步骤如下。

(1) 在 xml 文件布局上实现 ListView。

```
<?xml version="1.0" encoding="utf-8"?>
<RelativeLayoutxmlns:android="http://schemas.android.com/apk/res/android"
xmlns:tools="http://schemas.android.com/tools"
android:layout_width="match_parent"
android:layout_height="match_parent"
android:paddingBottom="@dimen/activity_vertical_margin"
android:paddingLeft="@dimen/activity_horizontal_margin"
android:paddingRight="@dimen/activity_horizontal_margin"
android:paddingTop="@dimen/activity_vertical_margin"
tools:context="com.example.carson_ho.adapte_demo.MainActivity">
```

```xml
<ListView
android:id="@+id/list_item"
android:layout_width="match_parent"
android:layout_height="match_parent"
android:divider="#f00"
android:dividerHeight="1sp"
android:headerDividersEnabled="false">
</ListView>
</RelativeLayout>
```

(2) 根据实际需求定制列表项：实现 ListView 每行的 xml 布局(即 item 布局)。

```xml
<?xml version="1.0" encoding="utf-8"?>
<RelativeLayoutxmlns:android="http://schemas.android.com/apk/res/android"
android:layout_width="match_parent"
android:layout_height="match_parent">

<TextView
android:id="@+id/name"
android:layout_width="wrap_content"
android:layout_height="wrap_content"
android:textSize="17sp"
android:paddingLeft="14dp"/>
<TextView
android:id="@+id/address"
android:layout_below="@id/name"
android:textSize="17sp"
android:layout_width="wrap_content"
android:layout_height="wrap_content" />
<TextView
android:id="@+id/lowerest_wholesale"
android:layout_toRightOf="@id/address"
android:textSize="17sp"
android:layout_width="wrap_content"
android:layout_height="wrap_content" />
<TextView
android:id="@+id/price"
android:textSize="17sp"
android:layout_below="@id/address"
android:layout_width="wrap_content"
android:layout_height="wrap_content" />
<ImageView
android:id="@+id/picture"
android:layout_alignParentRight="true"
android:layout_width="115dp"
android:layout_height="86dp"            />
</RelativeLayout>
```

(3) 定义一个 HashMap 构成的列表以键值对的方式存放数据。

(4) 构造 SimpleAdapter 对象，设置适配器。

(5) 将 LsitView 绑定到 SimpleAdapter 上。

```java
publicclassMainActivityextendsAppCompatActivity {
    //定义数组以填充数据
    private String[] name=new String[]{
        "威龙注塑机""霸龙注塑机""恐龙注塑机"     };
    private String[] address =new String[]{
        "广东""北京""黑龙江"     };
    privateint[] lowerest_wholesale =newint[]{
        11,22,33     };
    privateint[] price =newint[]{
        11,22,33     };
    privateint[] picture =newint[]{
        R.drawable.home_selected,
        R.drawable.home_selected,
        R.drawable.home_selected     };

    @Override
    protectedvoid onCreate(Bundle savedInstanceState) {
        super.onCreate(savedInstanceState);
        setContentView(R.layout.activity_main);

        //定义一个 HashMap 构成的列表以键值对的方式存放数据
        ArrayList<HashMap<String, Object>> listItem = new ArrayList<HashMap<String,Object>>();
        //循环填充数据
        for(int i=0;i<name.length;i++)         {
            HashMap<String,Object> map = new HashMap<String,Object>();
            map.put("name", name[i]);
            map.put("address", address[i]);
            map.put("lowerest_wholesale", lowerest_wholesale[i]);
            map.put("price", price[i]);
            map.put("picture", picture[i]);
            listItem.add(map);
        }

        //构造 SimpleAdapter 对象，设置适配器
        SimpleAdapter mSimpleAdapter = new SimpleAdapter(this,
            listItem,                    //需要绑定的数据
            R.layout.item_imformation,   //每一行的布局
            new String[] {"name","address", "lowerest_wholesale","price","picture"},
            //数组中的数据源的键对应到定义布局的 View 中
            newint[] {R.id.name,R.id.address,R.id.lowerest_wholesale,R.id.price,R.id.picture});
        ListView list= (ListView) findViewById(R.id.list_item);
        //为 ListView 绑定适配器
        list.setAdapter(mSimpleAdapter);
    }
}
```

结果如图 6-15 所示。

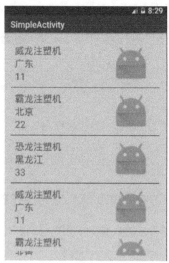

图 6-15

3. BaseAdapter 详解

BaseAdapter 是应用最多的一种适配器。它是一个抽象类，需要重写方法完成自定义适配器的功能，比较自由灵活，能实现各种想要的效果。

操作步骤如下。

(1) 定义主 xml 布局。

(2) 根据需要，定义 ListView 每行所实现的 xml 布局。

(3) 定义一个 Adapter 类继承 BaseAdapter，重写里面的方法。

(4) 定义一个 HashMap 构成的列表，将数据以键值对的方式存放在里面。

(5) 构造 Adapter 对象，设置适配器。

(6) 将 LsitView 绑定到 Adapter 上。

定义一个 Adapter 类继承 BaseAdapter，并重写里面的方法。使用 BaseAdapter 必须写一个类继承它，同时 BaseAdapter 是一个抽象类，继承它必须实现它的方法。

```
classMyAdapterextendsBaseAdapter{
    private LayoutInflater mInflater;//得到一个 LayoutInfalter 对象用来导入布局
    //构造函数
    publicMyAdapter(Context context,ArrayList<HashMap<String, Object>> listItem){
        this.mInflater = LayoutInflater.from(context);
        this.listItem = listItem;
    }//声明构造函数

    @Override
    publicintgetCount(){
        return listItem.size();
    }//这个方法返回了在适配器中所代表的数据集合的条目数
```

```
@Override
public Object getItem(int position){
    return listItem.get(position);
}//这个方法返回了数据集合中与指定索引 position 对应的数据项

@Override
publiclonggetItemId(int position){
    return position;
}//这个方法返回了在列表中与指定索引对应的行 id


@Override
public View getView(int position, View convertView, ViewGroup parent){
    returnnull;
}//这个方法返回了指定索引对应的数据项的视图，还没写完
}
```

这里主要讲的是 BaseAdapter 中必须要重写的 4 个方法。BaseAdapter 的灵活性就在于它要重写很多方法，其中最重要的即为 getView() 方法。

我们结合上述重写的 4 个方法了解 ListView 的绘制过程，如图 6-16 所示。

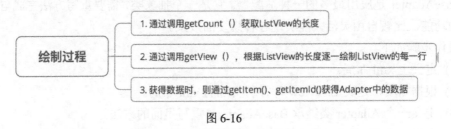

图 6-16

其中，重点讲解重写的 getView() 方法，总共有 3 种，代码如下。

```
/**
 * 重写方式1：直接返回指定索引对应的数据项的视图
 */
@Override
public View getView(int position, View convertView, ViewGroup parent){
    View item = mInflater.inflate(R.layout.item,null);
    ImageView img = (ImageView)item.findViewById(R.id.ItemImage);
    TextView title = (TextView)item.findViewById(R.id.ItemTitle);
    TextView test = (TextView)item.findViewById(R.id.ItemText);
    Button btn = (Button) item.findViewById(R.id.ItemBottom);
    img.setImageResource((Integer) listItem.get(position).get("ItemImage"));
    title.setText((String) listItem.get(position).get("ItemTitle"));
    test.setText((String) listItem.get(position).get("ItemText"));

    return item;
}
// 缺点：
```

```java
// 每次调用 getView()时，都要重新通过 findViewById()寻找 View 组件和重新绘制 View
// 当列表项数据量很大时会严重影响性能，即体现为下拉很慢、很卡

/**
 * 重写方式 2：使用 convertView 作为 View 缓存(优化)
 * 具体原理：
 *         // a. 将 convertView 作为 getView()的输入参数和返回参数，从而形成反馈
 *         // b. 形成了 Adapter 的 itemView 重用机制，减少了重绘 View 的次数
 */
@Override
public View getView(int position, View convertView, ViewGroup parent){

    // 检测有无可重用的 View，若无就重新绘制
    if(convertView == null)
    {
        convertView = mInflater.inflate(R.layout.item, null);
    }
    ImageView img = (ImageView)convertView.findViewById(R.id.ItemImage);
    TextView title = (TextView)convertView.findViewById(R.id.ItemTitle);
    TextView test = (TextView)convertView.findViewById(R.id.ItemText);
    Button btn = (Button) convertView.findViewById(R.id.ItemBottom);
    img.setImageResource((Integer) listItem.get(position).get("ItemImage"));
    title.setText((String) listItem.get(position).get("ItemTitle"));
    test.setText((String) listItem.get(position).get("ItemText"));

    return convertView;
    // 最终返回 convertView 形成反馈
}
// 优点：减少了重绘 View 的次数
// 缺点：每次都要通过 findViewById()寻找 View 组件
/*** 重写方式 3：在方式 2 的基础上，使用 ViewHolder 实现更加具体的缓存——View 组件缓存
 * 具体原理：
 *         // a. 将 convertView 作为 getView()的输入参数和返回参数，从而形成反馈
 *         // b. 形成了 Adapter 的 itemView 重用机制，减少了重绘 View 的次数
 */
staticclassViewHolder
{
    public ImageView img;
    public TextView title;
    public TextView text;
    public Button btn;
}
@Override
public View getView(int position, View convertView, ViewGroup parent){
    ViewHolder holder ;
    if(convertView == null)
    {
```

```
            holder = new ViewHolder();
            convertView = mInflater.inflate(R.layout.item, null);
            holder.img = (ImageView)convertView.findViewById(R.id.ItemImage);
            holder.title = (TextView)convertView.findViewById(R.id.ItemTitle);
            holder.text = (TextView)convertView.findViewById(R.id.ItemText);
            holder.btn = (Button) convertView.findViewById(R.id.ItemBottom);
            convertView.setTag(holder);
        }
        else {
            holder = (ViewHolder)convertView.getTag();
        }
            holder.img.setImageResource((Integer) listItem.get(position).get("ItemImage"));
            holder.title.setText((String) listItem.get(position).get("ItemTitle"));
            holder.text.setText((String) listItem.get(position).get("ItemText"));

        return convertView;
}
```

重用 View 时不用通过 findViewById()重新寻找 View 组件，同时也减少了重绘 View 的次数，是 ListView 使用的最优化方案。

方案 3(通过 convertView+ViewHolder 重写 getView())是 ListView 使用的最优化方案，所以非常推荐大家使用。

ListView 的优化，如图 6-17 所示。

图 6-17

方案 3 的完整实现操作步骤如下。
(1) 定义主 xml 的布局。
activity_main.xml：

```
<?xml version="1.0" encoding="utf-8"?>
<LinearLayoutxmlns:android="http://schemas.android.com/apk/res/android"
xmlns:tools="http://schemas.android.com/tools"
android:layout_width="match_parent"
android:layout_height="match_parent"
android:background="#FFFFFF"
android:orientation="vertical">
<ListView
```

```xml
    android:id="@+id/listView1"
    android:layout_width="match_parent"
    android:layout_height="match_parent" />
</LinearLayout>
```

(2) 根据需要，定义 ListView 每行所实现的 xml 布局(item 布局)。
item.xml：

```xml
<?xml version="1.0" encoding="utf-8"?>
<RelativeLayoutxmlns:android="http://schemas.android.com/apk/res/android"
android:layout_width="match_parent"
android:layout_height="match_parent">
<ImageView
android:layout_alignParentRight="true"
android:layout_width="wrap_content"
android:layout_height="wrap_content"
android:id="@+id/ItemImage"/>
<Button
android:layout_width="wrap_content"
android:layout_height="wrap_content"
android:text="按钮"
android:id="@+id/ItemBottom"
android:focusable="false"
android:layout_toLeftOf="@+id/ItemImage" />
<TextViewandroid:id="@+id/ItemTitle"
android:layout_height="wrap_content"
android:layout_width="fill_parent"
android:textSize="20sp"/>
<TextViewandroid:id="@+id/ItemText"
android:layout_height="wrap_content"
android:layout_width="fill_parent"
android:layout_below="@+id/ItemTitle"/>
</RelativeLayout>
```

(3) 定义一个 Adapter 类继承 BaseAdapter，重写里面的方法(利用 convertView+ ViewHolder 来重写 getView())。
MyAdapter.java：

```java
package scut.learnlistview;

import android.content.Context;
import android.view.LayoutInflater;
import android.view.View;
import android.view.ViewGroup;
import android.widget.BaseAdapter;
import android.widget.Button;
import android.widget.ImageView;
```

```java
import android.widget.TextView;
import java.util.ArrayList;
import java.util.HashMap;

/**
 * Created by yany on 2016/4/11.
 */
classMyAdapterextendsBaseAdapter{
private LayoutInflater mInflater;//得到一个 LayoutInfalter 对象用来导入布局
    ArrayList<HashMap<String, Object>> listItem;

publicMyAdapter(Context context,ArrayList<HashMap<String, Object>> listItem){
    this.mInflater = LayoutInflater.from(context);
    this.listItem = listItem;
    }//声明构造函数


@Override
publicintgetCount(){
    return listItem.size();
    }//这个方法返回了在适配器中所代表的数据集合的条目数


@Override
public Object getItem(int position){
    return listItem.get(position);
    }//这个方法返回了数据集合中与指定索引 position 对应的数据项


@Override
publiclonggetItemId(int position){
    return position;
    }//这个方法返回了在列表中与指定索引对应的行 id

//利用 convertView+ViewHolder 来重写 getView()
staticclassViewHolder
{
    public ImageView img;
    public TextView title;
    public TextView text;
    public Button btn;
    }//声明一个外部静态类
@Override
public View getView(finalint position, View convertView, final ViewGroup parent){
    ViewHolder holder ;
    if(convertView == null)
    {
        holder = new ViewHolder();
        convertView = mInflater.inflate(R.layout.item, null);
        holder.img = (ImageView)convertView.findViewById(R.id.ItemImage);
```

```java
            holder.title = (TextView)convertView.findViewById(R.id.ItemTitle);
            holder.text = (TextView)convertView.findViewById(R.id.ItemText);
            holder.btn = (Button) convertView.findViewById(R.id.ItemBottom);
            convertView.setTag(holder);
        }
    else {
            holder = (ViewHolder)convertView.getTag();

        }
            holder.img.setImageResource((Integer) listItem.get(position).get("ItemImage"));
            holder.title.setText((String) listItem.get(position).get("ItemTitle"));
            holder.text.setText((String) listItem.get(position).get("ItemText"));
            holder.btn.setOnClickListener(new View.OnClickListener() {
@Override
    publicvoidonClick(View v){
                System.out.println("你单击了选项"+position);//bottom 会覆盖 item 的焦点,所以要
                    在 xml 中配置 android:focusable="false"
            }
        });

        return convertView;
    }//这个方法返回了指定索引对应的数据项的视图
}
```

(4) 在 MainActivity 中：
- 定义一个 HashMap 构成的列表，将数据以键值对的方式存放在里面。
- 构造 Adapter 对象，设置适配器。
- 将 LsitView 绑定到 Adapter 上。

MainActivity.java：

```java
package scut.learnlistview;

import android.support.v7.app.AppCompatActivity;
import android.os.Bundle;
import android.view.View;
import android.widget.AdapterView;
import android.widget.ArrayAdapter;
import android.widget.ListView;
import android.widget.SimpleAdapter;
import java.util.ArrayList;
import java.util.HashMap;
import java.util.List;

publicclassMainActivityextendsAppCompatActivity{
    private ListView lv;

@Override
```

```java
publicvoidonCreate(Bundle savedInstanceState){
    super.onCreate(savedInstanceState);
    setContentView(R.layout.activity_main);

    lv = (ListView) findViewById(R.id.listView1);
    /*定义一个以 HashMap 为内容的动态数组*/
    ArrayList<HashMap<String, Object>> listItem = new ArrayList<HashMap<String, Object>>();
    /*在数组中存放数据*/
    for (int i = 0; i <100; i++) {
        HashMap<String, Object> map = new HashMap<String, Object>();
        map.put("ItemImage", R.mipmap.ic_launcher);//加入图片
        map.put("ItemTitle", "第" + i + "行");
        map.put("ItemText", "这是第" + i + "行");
        listItem.add(map);
    }
    MyAdapter adapter = new MyAdapter(this, listItem);
    lv.setAdapter(adapter);//为 ListView 绑定适配器

    lv.setOnItemClickListener(new AdapterView.OnItemClickListener() {
        @Override
        publicvoidonItemClick(AdapterView<?> arg0, View arg1, int arg2, long arg3){
            System.out.println("你单击了第" + arg2 + "行");//设置系统输出单击的行
        }
    });
}
```

运行结果如图 6-18 所示，输出结果如图 6-19 所示。

图 6-18   图 6-19

6.6 GridView

　　Android 中的网格视图(GridView)主要是把一系列空间组织成一个二维的网格显示出来。当屏幕上是以图标作为应用程序的入口，特别是图标比较多时，我们可以考虑使用 GridView 来进行布局，这样用户使用起来比较直观，操作方便。原理上除了可以设置多列展示之外，其他与 ListView 基本相同。

　　GridView 的常用属性及描述如表 6-3 所示。

表 6-3

属性	描述
android:columnWidth	设置列的宽度
android:gravity	设置此组件中的内容在组件中的位置。可选的值有 top、bottom、left、right、center_vertical、fill_vertical、center_horizontal、fill_horizontal、center、fill、clip_vertical，可以多选，用"|"分开
android:horizontalSpacing	两列之间的间距
android:numColumns	设置列数
android:stretchMode	缩放模式
android:verticalSpacing	两行之间的间距

　　示例效果如图 6-20 所示。

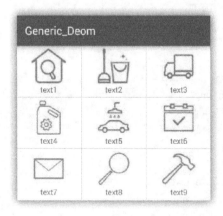

图 6-20

　　操作步骤如下。
　　(1) 布局 activity_main。

```
<?xml version="1.0" encoding="utf-8"?>
<LinearLayout xmlns:android="http://schemas.android.com/apk/res/android"
    xmlns:tools="http://schemas.android.com/tools"
    android:layout_width="match_parent"
    android:layout_height="match_parent"
```

```
        tools:context="custom.community.com.generic_deom.MainActivity">

<GridView
    android:id="@+id/gridview"
    android:layout_width="fill_parent"
    android:layout_height="wrap_content"

    android:listSelector="@null"
    android:horizontalSpacing="0.5dp"
    android:verticalSpacing="0.5dp"
    android:background="#c9c9c9"
    android:numColumns="3"
    android:scrollbars="none"
    android:stretchMode="columnWidth"

    />

</LinearLayout>
```

(2) item 布局。

```
<?xml version="1.0" encoding="utf-8"?>
<LinearLayout xmlns:android="http://schemas.android.com/apk/res/android"
    android:orientation="vertical"
    android:layout_width="match_parent"
    android:layout_height="match_parent"
    android:background="#ffffff"
    android:gravity="center"
    android:padding="4dp"
    >
<ImageView
    android:id="@+id/item_img"
    android:layout_width="64dp"
    android:layout_height="64dp"
    />
<TextView
    android:id="@+id/item_text"
    android:layout_width="wrap_content"
    android:layout_height="wrap_content"
    />
</LinearLayout>
```

(3) MainActivity 代码。

```
public class MainActivity extends AppCompatActivity {
    private GridView gridview;

    GridViewSim myGridView;
    LayoutInflater   inflater;
```

```java
//定义两个数组，放图片和文字
private String text[]={"text1","text2","text3","text4","text5","text6","text7","text8","text9"};
private int img_grid[] ={R.drawable.a,R.drawable.b,R.drawable.c,
    R.drawable.d,R.drawable.e,R.drawable.f,R.drawable.g,R.drawable.h,R.drawable.i};
@Override
protected void onCreate(Bundle savedInstanceState) {
    super.onCreate(savedInstanceState);
    setContentView(R.layout.activity_main);

    gridview=(GridView)findViewById(R.id.gridview);

    myGridView=new GridViewSim(this,text,img_grid);
    gridview.setAdapter(myGridView);

    inflater= LayoutInflater.from(this);
}


class GridViewSim extends BaseAdapter{
    private Context context=null;
    private String data[]=null;
    private int imgId[]=null;


    private class Holder{

        ImageView item_img;
        TextView item_tex;

        public ImageView getItem_img() {
            return item_img;
        }

        public void setItem_img(ImageView item_img) {
            this.item_img = item_img;
        }

        public TextView getItem_tex() {
            return item_tex;
        }

        public void setItem_tex(TextView item_tex) {
            this.item_tex = item_tex;
        }
```

```java
    }
    //构造方法
    public GridViewSim(Context context, String[] data, int[] imgId) {
        this.context = context;
        this.data = data;
        this.imgId = imgId;
    }


    @Override
    public int getCount() {


        return data.length;

    }

    @Override
    public Object getItem(int position) {
        return position;
    }

    @Override
    public long getItemId(int position) {
        return position;
    }

    @Override
    public View getView(int position, View view, ViewGroup viewGroup) {
        Holder holder;
          if(view==null){
            view=inflater.inflate(R.layout.item_listview,null);
             holder=new Holder();
              holder.item_img=(ImageView)view.findViewById(R.id.item_img);
              holder.item_tex=(TextView)view.findViewById(R.id.item_text);
              view.setTag(holder);
          }else{
              holder=(Holder) view.getTag();
          }
        holder.item_tex.setText(data[position]);
        holder.item_img.setImageResource(imgId[position]);

        return view;
    }
}
}
```

6.7 ProgressBar

当一个应用程序在后台执行时，前台界面就不会有信息，此时用户根本不知道程序是否在运行、运行进度如何、应用程序是否遇到错误终止，这时需要使用进度条来提示用户后台程序执行的进度。Android 系统提供了两大类进度条样式，即长形进度条和圆形进度条，如图 6-21 所示。

具体实现如下。

activity_main.xml：

```xml
<RelativeLayout xmlns:android="http://schemas.android.com/apk/res/android"
    xmlns:tools="http://schemas.android.com/tools"
    android:layout_width="fill_parent"
    android:layout_height="fill_parent"
    tools:context=".MainActivity" >
<LinearLayout
        android:layout_width="fill_parent"
        android:layout_height="fill_parent"
        android:orientation="vertical" >
<ProgressBar
        android:id="@+id/bar1"
        style="?android:attr/progressBarStyleHorizontal"
        android:layout_width="300dip"
        android:layout_height="wrap_content"
        />
<ProgressBar
        android:id="@+id/bar2"
        style="?android:attr/progressBarStyleLarge"
        android:layout_width="wrap_content"
        android:layout_height="wrap_content"
        android:max="100"
        android:progress="50"
        android:secondaryProgress="70"
        />

<Button
        android:id="@+id/btok"
        android:layout_width="100dip"
        android:layout_height="wrap_content"
        android:text="@string/start" />
</LinearLayout>
</RelativeLayout>
```

图 6-21

MainActivity.java：

```java
package com.example.admin.myapplication;

import android.app.Dialog;
import android.app.ProgressDialog;
import android.content.DialogInterface;
import android.content.Intent;
import android.os.Handler;
import android.os.Message;
import android.support.v7.app.AlertDialog;
import android.support.v7.app.AppCompatActivity;
import android.os.Bundle;
import android.view.View;
import android.widget.AdapterView;
import android.widget.ArrayAdapter;
import android.widget.Button;
import android.widget.CheckBox;
import android.widget.CompoundButton;
import android.widget.ProgressBar;
import android.widget.Spinner;
import android.widget.Toast;

public class MainActivity extends AppCompatActivity implements View.OnClickListener {
    private int count = 0;
    private static final int GUI_STOP_NOTIFIER = 0x108;
    private static final int GUI_THERADING_NOTIFIER = 0x109;
    private ProgressBar bar1;
    private ProgressBar bar2;
    private Button btok;

    @Override
    protected void onCreate(Bundle savedInstanceState) {
        super.onCreate(savedInstanceState);
        setContentView(R.layout.activity_main);

        bar1 = (ProgressBar) this.findViewById(R.id.bar1);
        bar2 = (ProgressBar) this.findViewById(R.id.bar2);
        bar1.setIndeterminate(false);
        bar2.setIndeterminate(false);
        btok = (Button) this.findViewById(R.id.btok);
        btok.setOnClickListener(this);
    }
    @Override
    public void onClick(View v) {
        if (v.getId() == R.id.btok)
        {
            bar1.setVisibility(View.VISIBLE);
```

```java
            bar2.setVisibility(View.VISIBLE);
            bar1.setMax(100);
            bar1.setProgress(0);
            bar2.setProgress(0);
            new Thread(new Runnable()
            {
                @Override
                public void run()
                {
                    for (int i = 0; i < 100; i++)
                    {
                        try
                        {
                            count = (i + 1) * 20;
                            Thread.sleep(1000);
                            if (i == 4)
                            {
                                Message m = new Message();
                                m.what = MainActivity.GUI_STOP_NOTIFIER;
                                MainActivity.this.myMessageHandler.sendMessage(m);
                                break;
                            }
                            else
                            {
                                Message m = new Message();
                                m.what = MainActivity.GUI_THERADING_NOTIFIER;
                                MainActivity.this.myMessageHandler.sendMessage(m);
                            }
                        }
                        catch (InterruptedException e)
                        {
                            e.printStackTrace();
                        }
                    }
                }
            }).start();
        }
    }
    Handler myMessageHandler = new Handler()
    {
        public void handleMessage(Message msg)
        {
            switch (msg.what)
            {
                case MainActivity.GUI_STOP_NOTIFIER:
                    bar1.setVisibility(View.GONE);
                    bar2.setVisibility(View.GONE);
```

```
                    Thread.currentThread().interrupt();
                    break;
                case MainActivity.GUI_THERADING_NOTIFIER:
                    if (!Thread.currentThread().isInterrupted())
                    {
                        bar1.setProgress(count);
                        bar2.setProgress(count);
                        setProgress(count * 10);
                        setSecondaryProgress(count * 100);
                    }
                    break;
            }
        };
    };
}
```

【单元小结】

- 熟练操作单选框、多选框和下拉框。
- 掌握 ListView 和 GridView 的工作原理。
- 掌握 BaseAdapter 的工作原理。
- 熟悉 ListView 关联 Adapter 操作数据。

【单元自测】

1. 下列不是 AdapterView 子类的是(　　)。
   A. ListView　　　B. GridView　　　C. ScrollView　　　D. Spinner
2. 下列关于 ListView 适用的描述中，错误的是(　　)。
   A. 使用 ListView 必须适用 BaseAdapter 填充数据
   B. 使用 ListView，Activity 必须继承 ListViewActivity
   C. ListView 中每一项的视图布局既可以适用内置布局，也可以适用自定义布局
   D. ListView 中每一项被选中时，将会触发 ItemClick 事件
3. 在使用 RadioButton 时，要实现互斥选择需要用的组件是(　　)。
   A. ButtonGroup　　B. RadioButtons　　C. CheckBox　　D. RadioGroup
4. Android 中 ArrayAdapter 类是用于(　　)。
   A. 把数据绑定到组件上　　　　　　B. 把数据显示到 Activity 上
   C. 把数据传递给广播　　　　　　　D. 把数据传递给服务
5. 使进度条横向的属性是(　　)。
   A. @android:style/Widget.ProgressBar.Horizontal
   B. @android:style/ ProgressBar.Horizontal

C. @style/ Widget .ProgressBar.Horizontal

D. @style/ ProgressBar.Horizontal

【上机实战】

上机目标

- 熟练掌握 Android 比较复杂控件的开发。
- 使用 Android 开发学生注册页面。

上机练习

【问题描述】
使用 Android 常用控件与 Java 后台进行数据动态交互。

【问题分析】
使用 XML 完成表示层的工作，由 Java 完成后台数据操作。

在开发过程中，可以采取先确定 Activity_main.xml 中共有多少个控件，然后确定布局方式、线性布局，经过测试无误后，再开发底层。

【参考步骤】
Activity_main.xml：

```xml
<RelativeLayout xmlns:android="http://schemas.android.com/apk/res/android"
    xmlns:tools="http://schemas.android.com/tools"
    android:layout_width="match_parent"
    android:layout_height="match_parent"
    android:background="@drawable/environment"
    android:paddingBottom="@dimen/activity_vertical_margin"
    android:paddingLeft="@dimen/activity_horizontal_margin"
    android:paddingRight="@dimen/activity_horizontal_margin"
    android:paddingTop="@dimen/activity_vertical_margin"
    tools:context=".Reg" >
    <LinearLayout
        android:layout_width="fill_parent"
        android:layout_height="fill_parent"
        android:orientation="vertical" >
        <LinearLayout
            android:layout_width="fill_parent"
            android:layout_height="wrap_content"
            android:layout_marginTop="40dip"
            android:orientation="horizontal" >
            <TextView
```

```xml
            android:layout_width="wrap_content"
            android:layout_height="wrap_content"
            android:text="@string/myname" />
        <EditText
            android:id="@+id/regName"
            android:layout_width="200dip"
            android:layout_height="30dip"
            android:hint="@string/entername"
            android:textSize="14px" />
    </LinearLayout>
    <LinearLayout
        android:layout_width="fill_parent"
        android:layout_height="wrap_content"
        android:layout_marginTop="10dip"
        android:orientation="horizontal" >
        <TextView
            android:layout_width="wrap_content"
            android:layout_height="wrap_content"
            android:text="@string/mypsw" />
        <EditText
            android:id="@+id/regPsw"
            android:layout_width="200dip"
            android:layout_height="30dip"
            android:hint="@string/enterpsw"
            android:password="true"
            android:textSize="14px" />
    </LinearLayout>
    <LinearLayout
        android:layout_width="fill_parent"
        android:layout_height="wrap_content"
        android:layout_marginTop="10dip"
        android:orientation="horizontal" >
        <TextView
            android:layout_width="wrap_content"
            android:layout_height="wrap_content"
            android:text="@string/myclass" />
        <Spinner
            android:id="@+id/myClass"
            android:layout_width="150dip"
            android:layout_height="wrap_content"
            android:gravity="center_horizontal"/>
    </LinearLayout>
    <LinearLayout
        android:layout_width="fill_parent"
        android:layout_height="wrap_content"
        android:layout_marginTop="10dip"
        android:orientation="horizontal" >
```

```xml
            <TextView
                android:layout_width="wrap_content"
                android:layout_height="wrap_content"
                android:text="@string/mysex" />
            <RadioGroup
                android:id="@+id/myradioGroup"
                android:layout_width="wrap_content"
                android:layout_height="wrap_content"
                android:orientation="horizontal" >
                <RadioButton
                    android:id="@+id/rad1"
                    android:layout_width="wrap_content"
                    android:layout_height="wrap_content"
                    android:text="@string/sexMan" />
                <RadioButton
                    android:id="@+id/rad2"
                    android:layout_width="wrap_content"
                    android:layout_height="wrap_content"
                    android:text="@string/sexWoman" />
            </RadioGroup>
        </LinearLayout>
        <LinearLayout
            android:layout_width="fill_parent"
            android:layout_height="wrap_content"
            android:layout_marginTop="5dip"
            android:gravity="center_horizontal"
            android:orientation="horizontal" >
            <Button
                android:id="@+id/myok"
                android:layout_width="100dip"
                android:layout_height="wrap_content"
                android:text="@string/myok" />
        </LinearLayout>
    </LinearLayout>
</RelativeLayout>
```

Reg.java：

```java
package com.hpsvse.test;

import android.app.Activity;
import android.content.Intent;
import android.os.Bundle;
import android.view.Gravity;
import android.view.Menu;
import android.view.View;
import android.view.View.OnClickListener;
import android.view.Window;
```

```java
import android.view.WindowManager;
import android.widget.AdapterView;
import android.widget.AdapterView.OnItemSelectedListener;
import android.widget.ArrayAdapter;
import android.widget.Button;
import android.widget.EditText;
import android.widget.RadioButton;
import android.widget.RadioGroup;
import android.widget.RadioGroup.OnCheckedChangeListener;
import android.widget.Spinner;
import android.widget.Toast;
import com.hpsvse.dao.UserDAO;
import com.hpsvse.util.DBConnection;

public class Reg extends Activity implements OnClickListener,
        OnItemSelectedListener, OnCheckedChangeListener
{
    private EditText txt1, txt2;
    private Button btok;
    private Spinner myClass;
    private String[] allClass =
{ "计算机(1)班", "计算机(2)班", "计算机(3)班", "计算机(4)班" };
    private String messageSpin, messageRadio;
    private ArrayAdapter adapter;
    private RadioGroup myGroup;
    private RadioButton radMan, radWoman;

    @Override
    protected void onCreate(Bundle savedInstanceState)
    {
        super.onCreate(savedInstanceState);
        requestWindowFeature(Window.FEATURE_NO_TITLE);
        getWindow().setFlags(WindowManager.LayoutParams.FLAG_FULLSCREEN,
        WindowManager.LayoutPara ms.FLAG_FULLSCREEN);
        setContentView(R.layout.activity_reg);
        init();
    }

    public void init()
    {
        txt1 = (EditText) this.findViewById(R.id.regName);
        txt2 = (EditText) this.findViewById(R.id.regPsw);
        myClass = (Spinner) this.findViewById(R.id.myClass);
        adapter = new ArrayAdapter(this, android.R.layout.simple_spinner_item,
                allClass);
        adapter.setDropDownViewResource(android.R.layout.simple_spinner_dropdown_item);
        myClass.setAdapter(adapter);
```

```java
        myClass.setOnItemSelectedListener(this);
        myGroup = (RadioGroup) this.findViewById(R.id.myradioGroup);
        radMan = (RadioButton) this.findViewById(R.id.rad1);
        radWoman = (RadioButton) this.findViewById(R.id.rad2);
        myGroup.setOnCheckedChangeListener(this);
        btok = (Button) this.findViewById(R.id.myok);
        btok.setOnClickListener(this);
    }
    /***
     * 按钮事件
     */
    @Override
    public void onClick(View v)
    {
        String username = txt1.getText().toString();
        String psw = txt2.getText().toString();
        if (!"".equals(username) && !"".equals(psw))
        {
            // 存入数据库
        }
    }
    /***
     * 下拉框事件
     */
    @Override
    public void onItemSelected(AdapterView<?> arg0, View arg1, int arg2,
            long arg3)
    {
        messageSpin = allClass[arg2];
        arg0.setVisibility(arg1.VISIBLE);
    }

    @Override
    public void onNothingSelected(AdapterView<?> arg0)
    {
        // TODO Auto-generated method stub
    }

    /****
     * 单选按钮
     */
    @Override
    public void onCheckedChanged(RadioGroup group, int checkedId)
    {
        if (checkedId == radMan.getId())
        {
            messageRadio = radMan.getText().toString();
```

```
            }
            if (checkedId == radWoman.getId())
            {
                messageRadio = radWoman.getText().toString();
            }
        }
        public void myPlay(String str)
        {
            Toast t = Toast.makeText(this, str, Toast.LENGTH_SHORT);
            t.setGravity(Gravity.TOP, 0, 200);
            t.show();
        }
        @Override
        public boolean onCreateOptionsMenu(Menu menu)
        {
            getMenuInflater().inflate(R.menu.reg, menu);
            return true;
        }
    }
```

首页效果如图 6-22 所示。

图 6-22

【拓展作业】

1. 完成一个复杂的学生信息注册(自己设计 UI，尽可能多地使用控件)。
2. 请默写几个重点控件的事件函数的原型，如 ListView、GridView、ProgressBar 等。

单元七
Android 高级控件

 课程目标

- ▶ Toolbar
- ▶ ViewPager
- ▶ Fragment
- ▶ TabLayout 导航栏

简介

Toolbar 是谷歌在 2014 年 Google IO 大会上推出的一套全新的设计规范——Material Design。它的出现规范了 Android 开发者开发 APP 标题栏的设计风格，极大地提高了开发效率，而且 Material Design 设计规范也越来越多地出现在我们常用的 APP 中。

Android 3.0 中引入了 fragments 的概念，主要是用在大屏幕设备上，如平板电脑上，支持更加动态和灵活的 UI 设计。平板电脑屏幕比手机屏幕大得多，因此有更多的空间来放更多的 UI 组件，并且这些组件之间会产生更多的交互。

ViewPager 是 google SDK V4 包中自带的一个视图类，主要是用来实现多个屏幕间的切换。很多底层原生的界面也还是依然采用 ViewPager+Fragment 的布局方式，事实上这依然是主流，微信 5.0 到 6.0 等也还是采用这种布局方法，所以还是非常有必要学习一下。

7.1 Toolbar

7.1.1 了解 Toolbar

有关 Toolbar 的官方介绍中有一句话：A Toolbar is a generalization of action bars for use within application layouts. 意思是 Toolbar 是应用内的操作栏的一个归纳。这句话理解起来有些困难，我们只需要知道，Toolbar 是用来替代原来的 ActionBar 即可。下面我们来看一下 Toolbar 的类继承关系，如图 7-1 所示。

```
public class Toolbar
extends ViewGroup

java.lang.Object
   ↳ android.view.View
      ↳ android.view.ViewGroup
         ↳ androidx.appcompat.widget.Toolbar
```

图 7-1

从图 7-1 中的继承关系可以看出，Toolbar 继承的是 ViewGroup，即 Toolbar 是一个 ViewGroup 容器，知道了这一点对于后面的内容理解就比较容易了。下面我们从官方文档中看一下 ViewGroup 容器里的内容，如图 7-2 所示。

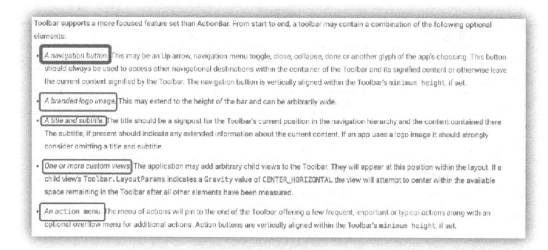

图 7-2

官方文档中显示，一个 Toolbar 从左到右包括 navigation button、logo、title、subtitle、自定义 View 和 action menu 5 个部分，即 ViewGroup 容器里包含了这 5 部分内容，如图 7-3 所示。

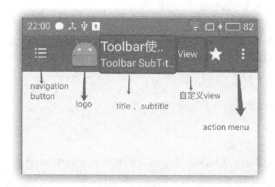

图 7-3

我们要为 ViewGroup 容器设置相应的内容(例如：设置了 navigation button，屏幕左上角导航按钮才会显示，否则，不显示，见 7.12 节详解)，否则它就只是一个空的 ViewGroup。至此，我们已了解了 Toolbar 的内容，下面我们介绍 Toolbar 的使用。

7.1.2 使用 Toolbar

首先看一下 Toolbar 有哪些比较常用和重要的方法。
Toolbar 常用的 XML 属性及说明如表 7-1 所示。

表 7-1

XML 属性	说明
toolbar:navigationIcon	设置 navigation button
toolbar:logo	设置 logo 图标
toolbar:title	设置标题
toolbar:titleTextColor	设置标题文字颜色
toolbar:subtitle	设置副标题
toolbar:subtitleTextColor	设置副标题文字颜色
toolbar:titleTextAppearance	设置 title text 相关属性，如字体、颜色、大小等
toolbar:subtitleTextAppearance	设置 subtitle text 相关属性，如字体、颜色、大小等
toolbar:logoDescription	logo 描述
android:background	Toolbar 背景
android:theme	主题

表 7-1 中的属性也可以在代码中设置，具体如下。

```
//设置 NavigationIcon
toolbar.setNavigationIcon(R.drawable.ic_book_list);
        // 设置 navigation button 单击事件
toolbar.setNavigationOnClickListener(new View.OnClickListener() {
            @Override
public void onClick(View v) {
finish();
        }
    });
        // 设置 Toolbar 背景色
toolbar.setBackgroundColor(getResources().getColor(R.color.colorPrimary));
        // 设置 Title
toolbar.setTitle(R.string.toolbar_title);
        // 设置 Toolbar title 文字颜色
toolbar.setTitleTextColor(getResources().getColor(R.color.white));
        // 设置 Toolbar subtitle
toolbar.setSubtitle(R.string.sub_title);

toolbar.setSubtitleTextColor(getResources().getColor(R.color.white));
        // 设置 logo
toolbar.setLogo(R.mipmap.ic_launcher);
        // 设置 NavigationIcon 单击事件
toolbar.setNavigationOnClickListener(new View.OnClickListener() {
            @Override
public void onClick(View v) {
finish();
        }
    });
```

```java
            //设置 Toolbar menu
toolbar.inflateMenu(R.menu.setting_menu);
            // 设置溢出菜单的图标
toolbar.setOverflowIcon(getResources().getDrawable(R.drawable.abc_ic_menu_moreoverflow_mtrl_alpha));
            // 设置 menu item 单击事件
toolbar.setOnMenuItemClickListener(new Toolbar.OnMenuItemClickListener() {
            @Override
publicbooleanonMenuItemClick(MenuItem item) {
switch (item.getItemId()){
caseR.id.item_setting:
                                //单击设置
break;
            }
return false;
            }
        });
```

为一个 Activity 添加 Toolbar 的操作步骤如下。

(1) 在 gradle 中添加 v7 appcompat 支持库。

```
compile 'com.android.support:appcompat-v7:24.2.1'
```

(2) 确保 Activity 是继承的 AppCompatActivity。

```java
public class ToolbarActivity extends AppCompatActivity{
}
```

(3) 在应用清单中，将<application>元素设置为使用 appcompat 的其中一个 NoActionBar 主题。使用这些主题中的一个可以防止应用使用原生 ActionBar 类提供应用栏。

```xml
<application      android:theme="@style/Theme.AppCompat.Light.NoActionBar" />
```

当然，也可不在application节点，而且在<activity>节点设置主题，如果每个activity都是一样的主题，则可以设置在application节点中。主题可选择任意一个appCompatNoActionBar主题或者它们的子主题。

(4) 在 Activity 的布局文件中添加 Toolbar。

```xml
<android.support.v7.widget.Toolbar
android:id="@+id/tool_bar_2"
android:layout_width="match_parent"
android:layout_height="wrap_content"
android:background="@color/colorPrimary"
toolbar:navigationIcon="@drawable/ic_book_list"
toolbar:title="@string/toolbar_title"
toolbar:titleTextColor="@color/white"
toolbar:theme="@style/ToolbarTheme"
toolbar:popupTheme="@style/ThemeOverlay.AppCompat.Light"
>
</android.support.v7.widget.Toolbar>
```

(5) 在 Activity 中对 Toolbar 做一些相关的操作，如设置标题、设置 navigation button 单击事件、添加溢出菜单。

```java
private void initToolbar(){
    Toolbar toolbar = (Toolbar) findViewById(R.id.tool_bar_2);
    toolbar.setNavigationOnClickListener(new View.OnClickListener() {
        @Override
        public void onClick(View v) {
            finish();
        }
    });
    //添加溢出菜单
    toolbar.inflateMenu(R.menu.setting_menu);
    // 添加菜单单击事件
    toolbar.setOnMenuItemClickListener(new Toolbar.OnMenuItemClickListener() {
        @Override
        publicbooleanonMenuItemClick(MenuItem item) {
            switch (item.getItemId()){
                caseR.id.item_setting:
                    //单击设置菜单
                    break;
            }
            return false;
        }
    });
}
```

(6) 溢出菜单文件如下。

```xml
<?xml version="1.0" encoding="utf-8"?>
<menu xmlns:android="http://schemas.android.com/apk/res/android"
xmlns:app="http://schemas.android.com/apk/res-auto">
<item android:id="@+id/item_collect"
android:icon="@drawable/ic_favorite_more"
android:title="收藏"
app:showAsAction="ifRoom"
    />

<item android:id="@+id/item_setting"
android:title="设置选项"
app:showAsAction="never"
    />
<item android:id="@+id/item_model"
android:title="夜间模式"
app:showAsAction="never"
    />
</menu>
```

注意，app:showAsAction属性是设置菜单的显示方法，取值共 5 种，主要应用的有ifRoom、never、always 3 种。ifRoom表示如果Toolbar 上有显示空间则显示在Toolbar 上，否则就展示在溢出菜单里；never 表示总是显示在溢出菜单里；always 表示总是显示在Toolbar 上。效果如图 7-4 所示。

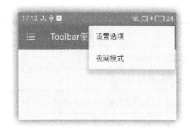

图 7-4

7.2 ViewPager 和 PagerAdapter

7.2.1 概述

ViewPager继承自ViewGroup，是左右两个屏幕平滑切换的一个容器，容器里呈现的视图由对应的Adapter决定，与其他标准的AdapterView类似。简而言之，我们通过Adapter把View放到ViewPager里，简单操作即可实现左右滑动互相切换View。另外，ViewPager的更新不是直接由其本身去完成的，而是通过观察者调用PagerAdapter的notifyDataSetChanged等相关方法去完成界面更新工作。

总的来看，ViewPager 负责显示页面、刷新页面、处理滑动等逻辑，而 PagerAdapter 负责实现如何渲染界面等具体接口。ViewPager 不直接操作页面，而是把一切逻辑都放在 PagerAdapter 里，甚至页面复用这些逻辑也交由 PagerAdapter 处理。

7.2.2 ViewPager 的重要属性

ViewPager 的常用 XML 属性及说明如表 7-2 所示。

表 7-2

XML 属性	说明
public ViewPager (Context context)	
public ViewPager (Context context, AttributeSetattrs)	
interface ViewPager.OnPageChangeListener	内部接口当切换不同 Page 时触发对应回调方法
void addView(Viewchild, int index, ViewGroup.LayoutParamsparams)	添加 View 到 ViewPager 中
boolean dispatchKeyEvent(KeyEventevent)	派发键盘事件到下一个视图
ViewGroup.LayoutParamsgenerateLayoutParams(AttributeSetattrs)	设置副标题文字颜色
PagerAdapter getAdapter()	获取对应的适配器

(续表)

XML 属性	说明
int getCurrentItem()	返回当前 page 的 index
int getPageMargin()	返回 Page 之间的外间距
boolean onTouchEvent(MotionEventev)	重写该方法可以处理 Touch 事件
void removeView(View view)	注意不能在 draw(android.graphics.Canvas)、onDraw(android.graphics.Canvas)、dispatchDraw(android.graphics.Canvas) 或其他相关方法中调用
void setPageTransformer(booleanreverseDrawingOrder, ViewPager.PageTransformer transformer)	当 the scroll position 改变时触发，可以用于设置自定义的动画效果，即只要实现 PageTransformer 接口和其唯一的方法 transformPage(View view, float position)
voidsetOnPageChangeListener(ViewPager.OnPageChangeListener listener)	设置 Page 切换事件监听
void setPageMarginDrawable(intresId)	
void setAdapter(PagerAdapter adapter)	设置 Adapter

7.2.3 PagerAdapter

我们先来介绍最普通的 PagerAdapter，如果想使用 PagerAdapter，则需要重写下面 4 个方法，其中 instantiateItem()方法和 isViewFromObject()方法涉及 key 的内容。当然，这只是官方建议，实际上我们只需重写 getCount()和 isViewFromObject()方法即可。

- getCount()：获得 viewpager 中有多少个 view。
- destroyItem()：移除一个给定位置的页面，适配器有责任从容器中删除当前操作的这个视图，即需要将 Position 对应的 view 从 container 中移除。这是为了确保在 finishUpdate(viewGroup)返回时视图能够被移除。
- instantiateItem()：①将给定位置的 view 添加到 ViewGroup(容器)中，创建并显示出来。②返回一个代表新增页面的 Object(key)，通常直接返回 view 本身即可，我们也可以定义自己的 key，但是 key 与每个 view 要是一一对应的关系。
- isViewFromObject()：判断 instantiateItem(ViewGroup, int)函数所返回来的 key 与一个页面视图是否代表的是同一个视图(即它们是否是对应的，若对应，则表示是同一个 View)，通常我们直接写为 return view == object！。

操作步骤如下。

(1) 编写 3 个内容相同的 view 视图，名字分别是 view_one、view_two、view_three。

view_one.xml：

<?xml version="1.0" encoding="utf-8"?>

```xml
<LinearLayoutxmlns:android="http://schemas.android.com/apk/res/android"
    android:layout_width="match_parent"
    android:layout_height="match_parent"
    android:background="#FFBA55"
    android:gravity="center"
    android:orientation="vertical">
    <TextView
        android:layout_width="wrap_content"
        android:layout_height="wrap_content"
        android:text="第一个 Page"
        android:textColor="#000000"
        android:textSize="18sp"
        android:textStyle="bold" />
</LinearLayout>
```

(2) 编写一个自定义的 PagerAdapter。

MyPagerAdapter.java：

```java
public class MyPagerAdapter extends PagerAdapter {
    privateArrayList<View>viewLists;
    publicMyPagerAdapter() {
    }

    publicMyPagerAdapter(ArrayList<View>viewLists) {
        super();
        this.viewLists = viewLists;
    }

    @Override
    publicintgetCount() {
        returnviewLists.size();
    }

    @Override
    publicbooleanisViewFromObject(View view, Object object) {
        return view == object;
    }

    @Override
    public Object instantiateItem(ViewGroup container, int position) {
        container.addView(viewLists.get(position));
        returnviewLists.get(position);
    }

    @Override
    public void destroyItem(ViewGroup container, int position, Object object) {
        container.removeView(viewLists.get(position));
    }
}
```

(3) 创建 Activity。

OneActivity.java：

```java
public class OneActivity extends AppCompatActivity{

    privateViewPagervpager_one;
    privateArrayList<View>aList;
    privateMyPagerAdaptermAdapter;

    @Override
    protected void onCreate(Bundle savedInstanceState) {
        super.onCreate(savedInstanceState);
        setContentView(R.layout.activity_one);
        vpager_one = (ViewPager) findViewById(R.id.vpager_one);

        aList = new ArrayList<View>();
        LayoutInflater li = getLayoutInflater();
        aList.add(li.inflate(R.layout.view_one,null,false));
        aList.add(li.inflate(R.layout.view_two,null,false));
        aList.add(li.inflate(R.layout.view_three,null,false));
        mAdapter = new MyPagerAdapter(aList);
        vpager_one.setAdapter(mAdapter);
    }
}
```

7.2.4 FragmentPagerAdapter 和 FragmentStatePagerAdapter

除 PagerAdapter 之外，Google 官方还建议我们使用 Fragment 来填充 ViewPager，这样可以更加方便地生成每个 Page，以及管理每个 Page 的生命周期。Fragment 给我们提供了两个专用的 Adapter，即 FragmentPagerAdapter 和 FragmentStatePagerAdapter，下面简要分析一下这两个 Adapter 的区别。

- FragmentPagerAdapter：与 PagerAdapter 一样，只能缓存当前的及左右各一个，共 3 个 Fragment，假如有 1、2、3、4 四个页面，则缓存如下。
  - 处于 1 页面：缓存 1、2。
  - 处于 2 页面：缓存 1、2、3。
  - 处于 3 页面：销毁 1 页面，缓存 2、3、4。
  - 处于 4 页面：销毁 2 页面，缓存 3、4。

  更多页面的情况，依次类推。

- FragmentStatePagerAdapter：当 Fragment 对用户不可见时，整个 Fragment 会被销毁，只保存 Fragment 的状态；而当页面需要重新显示时，会生成新的页面。

综上，FragmentPagerAdapter 适合固定页面较少的场合；而 FragmentStatePagerAdapter 则适合页面较多或页面内容非常复杂(需占用大量内存)的情况。

FragmentPagerAdapter 比较常用的方法有如下两种。
- getCount()：返回的是 ViewPager 页面的数量，即 Fragment 的数量。
- getItem(int position)：返回的是要显示的 Fragment 对象。

```
//实现一个FragmentPagerAdapter 至少需要实现 getCount()和 getItem()方法
FragmentPagerAdapter   fragmentadapter = new FragmentPagerAdapter(
getSupportFragmentManager()) {
    @Override
    publicintgetCount() {
        return fragments.size();   //getCount()方法返回的是 ViewPager 页面的数量，多少个
                                    Fragment 对应多少个页面
    }
    @Override
    public Fragment getItem(int position) {
        return fragments.get(position);//返回的是要显示的 fragment 对象
    }
}
};
```

7.3 Fragment

7.3.1 Fragment 概述

Fragment 是 Android 3.0 后引入的一个新的 API，它的出现是为了适应大屏幕的平板电脑，当然现在它仍然是平板 APP UI 设计的宠儿，并且现在普通手机的开发也会加入 Fragment。试想一下，如果有一个很大的界面，而我们只有一个布局，那么写起界面来会有多麻烦可想而知，而且如果组件很多，管理起来也很麻烦，而使用 Fragment 可以把屏幕划分成几块，然后进行分组，进行一个模块化的管理，从而可以更加方便地在运行过程中动态地更新 Activity 的用户界面。图 7-5 所示是文档中给出的一个 Fragment 分别对应平板电脑与手机的不同处理。

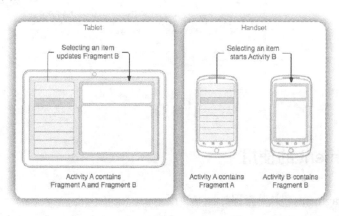

图 7-5

7.3.2 Fragment 的生命周期图

Fragment 的生命周期图如图 7-6 所示。

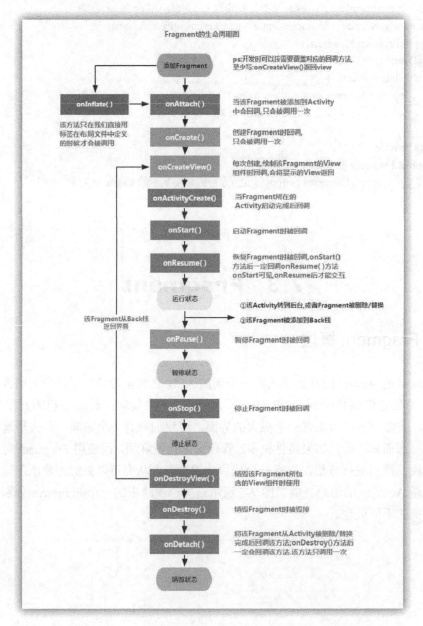

图 7-6

7.3.3 Fragment 的使用

静态加载 Fragment，实现流程如图 7-7 所示。

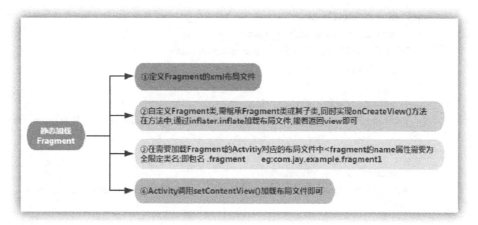

图 7-7

操作步骤如下。

(1) 创建 fragment1.xml 文件。

(2) 自定义一个 Fragment 类，需要继承 Fragment 或者它的子类，重写 onCreateView() 方法，在该方法中通过 inflater.inflate() 方法加载 fragment1.xml 文件，然后返回加载的 view 对象。

```
public class Fragmentone extends Fragment {
    @Override
    public View onCreateView(LayoutInflater inflater, ViewGroup container,
    Bundle savedInstanceState) {
        View view = inflater.inflate(R.layout.fragment1, container,false);
        return view;
    }
}
```

(3) 在需要加载 Fragment 的 Activity 对应的布局文件中添加 fragment 标签，注意，name 属性是全限定类名，即要包含 Fragment 的包名，代码如下。

```
<fragment
    android:id="@+id/fragment1"
    android:name="com.jay.example.fragmentdemo.Fragmentone"
    android:layout_width="match_parent"
    android:layout_height="0dp"
    android:layout_weight="1" />
```

(4) Activity 在 onCreate() 方法中调用 setContentView() 方法加载布局文件。

动态加载 Fragment，实现流程如图 7-8 所示。

Fragment 及布局代码此处不再展示，只展示 MainActivity 的关键代码，如下所示。

```
public class MainActivity extends Activity {
    @Override
    protected void onCreate(Bundle savedInstanceState) {
```

```
super.onCreate(savedInstanceState);
setContentView(R.layout.activity_main);
Display dis = getWindowManager().getDefaultDisplay();
if(dis.getWidth() > dis.getHeight())
{
    Fragment1 f1 = new Fragment1();
    getFragmentManager().beginTransaction().replace(R.id.LinearLayout1, f1).commit();
}
else{
    Fragment2 f2 = new Fragment2();
    getFragmentManager().beginTransaction().replace(R.id.LinearLayout1, f2).commit();
}
}
```

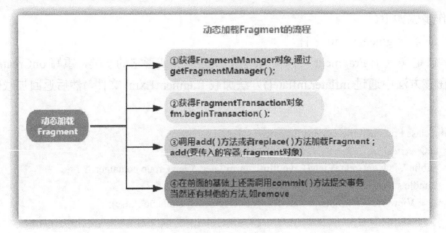

图 7-8

7.3.4　Fragment 管理与 Fragment 事务

Fragment 管理与 Fragment 事务流程如图 7-9 所示。

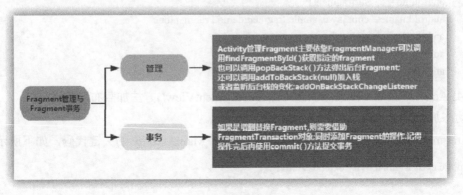

图 7-9

7.3.5 Fragment 与 Activity 的交互

1. 组件获取

首先 Fragment 可以通过 getActivity()方法来获得 Activity 的实例，然后 Activity 实例可以调用 findViewById()等方法拿到相应 id 的组件，这里组件的获取目的就是操作 Fragment 时达到可以修改 Activity 中组件的效果。

另外，在 Activity 中也可以获得一个 Fragment 的引用，从而调用 Fragment 中的方法并操作 Fragment 中的组件。获取方式如下：

getFragmentManager.findFragmentByid(R.id.fragment1)

2. 数据传递

1) Activity 传递数据给 Fragment

在 Activity 中创建 Bundle 数据包，调用 Fragment 实例的 setArguments(bundle)方法，从而将 Bundle 数据包传给 Fragment，然后在 Fragment 中调用 getArguments 获得 Bundle 对象，进而进行解析即可。

2) Fragment 传递数据给 Activity(接口方式)

(1) 在 Fragment 中准备回调接口，接口中声明传值的回调方法。

(2) 在 Fragment 中定义变量 private MyListener myListener。

(3) 重写 Fragment 中的 onAttach()方法：listener = (MyLisener)getActivity();。

(4) 在 Fragment 触发事件时回传值。

(5) 在 Activity 中实现回调接口，重写回调方法获取回传的值并显示。

Fragment 代码如下：

```java
public class MyFragment extends Fragment {

    private Button btn;
    private TextView text;
    private String str;

    //定义回调接口
    public interface MyListener{
        public void sendValue(String value);
    }

    private MyListener myListener;

    @Nullable
    @Override
    public View onCreateView(LayoutInflater inflater, @Nullable ViewGroup container, @Nullable Bundle savedInstanceState) {
        View v=inflater.inflate(R.layout.fragment,null);
```

```
            // 给 fragment 上的按钮添加单击事件
            text = v.findViewById(R.id.tv_value);
            btn = v.findViewById(R.id.btn_send);
        btn.setOnClickListener(new View.OnClickListener() {
            @Override
            public void onClick(View v) {
                //通过调用在 activity 中实现的接口方法，把数据传给 Mainactivity
                myListener.sendValue("传值");
            }
        });
            return v;
    }

    //activity 和 fragment 联系时调用，fragment 必须依赖 activity
        @Override
        public void onAttach(Context context) {
            super.onAttach(context);
            //获取实现接口的 activity
            myListener = (MyListener) getActivity();//或者 myListener=(MainActivity) context;

        }
    }
```

Activity 代码如下：

```
public class MainActivity extends BaseActivity implements MyListener{
//myFragment 中的接口实现
    public void sendValue(String value) {
        Log.e(TAG,value);
    }
}
```

7.4 TabLayout 导航栏

图 7-10 所示是某 Android 端应用的导航栏，此效果在其他 APP 的导航栏中也很常见，它的实现方式有很多，如 TabHost、自定义控件(第三方库)、RadioGroup 等。这里主要介绍 Android Design 库中 TabLayout 的使用。TabLayout 继承自 HorizontalScrollView，用作页面切换指示器，因使用简便、功能强大而广泛应用在 APP 中。

图 7-10

7.4.1 TabLayout 的常用属性

TabLayout 中常用的 XML 属性及说明如表 7-3 所示。

表 7-3

XML 属性	说明
app:tabIndicatorColor	指示线的颜色
app:tabIndicatorHeight	指示线的高度
app:tabSelectedTextColor	导航栏选中时的字体颜色
app:tabMode="scrollable"	可滚动的，默认是 fixed(固定的)

7.4.2 各种使用场景

TabLayout 属性有以下几个使用场景。

(1) 默认使用样式，结合Viewpager使用，效果如图 7-11 所示。

activity_main.xml 布局，代码如下。

图 7-11

```
<LinearLayout
    xmlns:android="http://schemas.android.com/apk/res/android"
    xmlns:tools="http://schemas.android.com/tools"
    android:layout_width="match_parent"
    android:layout_height="match_parent"
    android:orientation="vertical"
    tools:context="com.example.tablayoutusecase.
        defaultuse.MainActivity">

    <RelativeLayout
        android:layout_width="match_parent"
        android:layout_height="45dp"
        android:background=
            "@color/colorPrimaryDark">

        <TextView
            android:layout_width="wrap_content"
            android:layout_height="wrap_content"
            android:layout_centerInParent="true"
            android:text="一般用法"
            android:textColor="#fff"
            android:textSize="16sp"/>

    </RelativeLayout>
```

```xml
<android.support.design.widget.TabLayout
    android:id="@+id/tablayout"
    android:layout_width="match_parent"
    android:layout_height="wrap_content"/>


<android.support.v4.view.ViewPager
    android:id="@+id/viewpager"
    android:layout_width="match_parent"
    android:layout_height="match_parent"/>
</LinearLayout>
```

MainActivity 使用：根据 title 长度，设置文字 title、Fragment、Viewpager 联动，使用的是 TabLayout 默认属性。

```java
import android.support.design.widget.TabLayout;
import android.support.v4.app.Fragment;
import android.support.v4.view.ViewPager;
import android.support.v7.app.AppCompatActivity;
import android.os.Bundle;
import com.example.tablayoutusecase.R;
import java.util.ArrayList;

public class MainActivity extends AppCompatActivity {
    private TabLayout tabLayout;
    private ViewPager viewPager;
    private FmPagerAdapter pagerAdapter;
    private ArrayList<Fragment> fragments = new ArrayList<>();
    private String[] titles = new String[]{"最新","热门","我的"};

    @Override
    protected void onCreate(Bundle savedInstanceState) {
        super.onCreate(savedInstanceState);
        setContentView(R.layout.activity_main);

        init();
    }

    private void init() {

        tabLayout = (TabLayout) findViewById(R.id.tablayout);
        viewPager = (ViewPager) findViewById(R.id.viewpager);

        for(int i=0;i<titles.length;i++){
            fragments.add(new TabFragment());
            tabLayout.addTab(tabLayout.newTab());
        }
```

```
    tabLayout.setupWithViewPager(viewPager,false);
    pagerAdapter = new FmPagerAdapter(fragments,getSupportFragmentManager());
    viewPager.setAdapter(pagerAdapter);

    for(int i=0;i<titles.length;i++){
        tabLayout.getTabAt(i).setText(titles[i]);
    }
  }
}
```

(2) 设置 TabLayout 属性，效果如图 7-12 所示。

图 7-12

用 TabLayout 属性写一个 style，给需要的 TabLayout 引用。

```
<style name="MyTablayoutstyle" parent="Base.Widget.Design.TabLayout">
    <item name="tabBackground">@color/white</item>
    <item name="tabIndicatorColor">@color/green</item>
    <item name="tabIndicatorHeight">2dp</item>
    <item name="tabSelectedTextColor">@color/green</item>
    <item name="android:textSize">15sp</item>
    <item name="android:textColor">@color/text</item>
</style>
```

TabLayout 引用，代码如下。

```
<android.support.design.widget.TabLayout
    android:id="@+id/tab1"
    style="@style/MyTablayoutstyle"
    android:layout_width="match_parent"
    android:layout_height="wrap_content"/>
```

(3) TabLayout 去掉指示线，效果如图 7-13 所示。

图 7-13

这个操作很简单，将 tabIndicatorHeight 属性设置为 0dp，或者将 tabSelectedTextColor 属性设置为透明，即可隐藏指示线。

(4) 设置默认图标，效果如图 7-14 所示。

TabLayout 自带了 setIcon()方法设置图标资源，但这种效果不好，图容易被拉变形。可以自己造一个 tab 样式，如 tabLayout.getTabAt(i).setText(titles[i]).setIcon(pics[i])。效果如

图 7-15 所示。

图 7-14

图 7-15

创建图标和文字布局，代码如下。

```
<LinearLayout xmlns:android="http://schemas.android.com/apk/res/android"
    android:orientation="horizontal"
    android:layout_width="wrap_content"
    android:layout_height="48dp"
    android:gravity="center">

    <ImageView
        android:id="@+id/imageview"
        android:layout_gravity="center"
        android:layout_width="24dp"
        android:layout_height="24dp" />

    <TextView
        android:id="@+id/textview"
        android:layout_width="wrap_content"
        android:layout_height="match_parent"
        android:gravity="center"
        android:textSize="14sp"
        android:layout_marginLeft="8dp"/>
</LinearLayout>
```

(5) tab 数量太多，会超出屏幕，挤在一起。可通过 app:tabMode="scrollable"属性设置，让 TabLayout 变得可滚动、可超出屏幕。

效果如图 7-16 所示。

图 7-16

布局引入，代码如下。

```
<android.support.design.widget.TabLayout
    android:id="@+id/tablayout3"
    android:layout_width="match_parent"
    android:layout_height="wrap_content"
```

```
app:tabSelectedTextColor="@color/green"
android:layout_marginTop="20dp"
app:tabMode="scrollable"
android:background="@color/white"/>
```

【单元小结】

- 熟练使用 Toolbar 创建标题栏。
- 熟练使用 Viewpager+Fragment 做碎片化页面处理。
- 理解并搭建组合项目界面。

【单元自测】

1. Toolbar 中设置副标题的属性是(　　)。
   A. toolbar:logo　　　　　　　B. toolbar:title
   C. toolbar:titleTextColor　　　D. toolbar:subtitle
2. 下列不是 PagerAdapter 中的方法的是(　　)。
   A. getCount()　　　　　　　　B. destroyItem()
   C. instantiateItem()　　　　　 D. int getCurrentItem()
3. TabLayout 中 tab 数量太多，超出屏幕，若要显示完整，则需(　　)。
   A. 设置 tabIndicatorHeight =0dp　　B. 设置 app:tabMode="scrollable"
   C. 调用 setIcon()　　　　　　　　　D 设置 android:textSize
4. ViewPager 页面滚动的监听事件是(　　)。
   A. addView()　　　　　　　　　　B. onTouchEvent ()
   C. OnPageChangeListener()　　　　D. setPageMarginDrawable()

【上机实战】

上机目标

- 理解 ViewPager+Fragment 切换碎片化页面。
- 熟练操作 TabLayout+ViewPager+Fragment。

上机练习

图 7-17

【问题描述】

通过 TabLayout+ViewPager+Fragment 搭建一个企业级项目界面，如图 7-17 所示。

【问题分析】

页面最上方是 Toolbar 菜单栏；Toolbar 下面是 TabLayout 实现的导航栏；TabLayout 下面是 ViewPager 实现的切换 Fragment 页面。

【参考步骤】

实现图中效果，使用 TabLayout+ViewPager+Fragment。

(1) 布局。

```xml
<?xml version="1.0" encoding="utf-8"?>
  <com.zhy.autolayout.AutoLinearLayout xmlns:android="http://schemas.android.com/apk/res/android"
    xmlns:tools="http://schemas.android.com/tools"
     xmlns:app="http://schemas.android.com/apk/res-auto"
     android:layout_width="match_parent"
    android:layout_height="match_parent"
    android:orientation="vertical"
    android:background="@color/gray_line">
<include layout="@layout/item_title_haveback"/>
<android.support.design.widget.TabLayout
    android:id="@+id/tl_coupon"
    android:layout_width="match_parent"
    app:tabIndicatorColor="@color/golden"
    app:tabSelectedTextColor="@color/golden"
    app:tabTextColor="@color/txt_color_tint_"
    android:layout_height="90px">
</android.support.design.widget.TabLayout>
<android.support.v4.view.ViewPager
    android:id="@+id/vp_coupon"
    android:layout_width="match_parent"
    android:layout_height="match_parent"/>
</com.zhy.autolayout.AutoLinearLayout>
```

(2) Activity 实现。

```java
    ArrayList<String> titleDatas=new ArrayList<>();
        titleDatas.add("体验券");
        titleDatas.add("优惠券");
        ArrayList<Fragment> fragmentList = new ArrayList<Fragment>();
        fragmentList.add(new ExperienceFragment());
        fragmentList.add(new DiscountFragment());
        MyViewPageAdapter myViewPageAdapter = new
MyViewPageAdapter(getSupportFragmentManager(), titleDatas,         fragmentList);
        vpCoupon.setAdapter(myViewPageAdapter);
        tlCoupon.setupWithViewPager(vpCoupon);
        tlCoupon.setTabsFromPagerAdapter(myViewPageAdapter);
```

(3) MyViewPagerAdapter 实现(Fragment 创建省略)。

```java
public class MyViewPagerAdapter extends FragmentPagerAdapter {
    private ArrayList<String> titleList;
```

```java
        private ArrayList<Fragment> fragmentList;
        public MyViewPagerAdapter(FragmentManager fm, ArrayList<String> titleList,
            ArrayList<Fragment> fragmentList) {
            super(fm);
            this.titleList = titleList;
            this.fragmentList = fragmentList;
        }
        @Override
        public Fragment getItem(int position) {
            return fragmentList.get(position);
        }
        @Override
        public int getCount() {
            return fragmentList.size();
        }
        @Override
        public CharSequence getPageTitle(int position) {
            return titleList.get(position);
        }
    }
```

TabLayout+ViewPager 也可以实现布局。

(1) Activity 实现(布局同上)。

```java
private ArrayList<View> viewList = new ArrayList<>();
private TabLayout tlMain;
private ViewPager vpMain;
private MyAdapter adapter;

@Override
protected void onCreate(Bundle savedInstanceState) {
    super.onCreate(savedInstanceState);
    setContentView(R.layout.activity_main);
ArrayList<String> titleDatas=new ArrayList<>();
    titleDatas.add("体验券");
    titleDatas.add("优惠券");
    View view1 = getLayoutInflater().inflate(R.layout.view1, null);
    View view2 = getLayoutInflater().inflate(R.layout.view2, null);
    viewList.add(view1);
    viewList.add(view2);
    initView();
}

private void initView() {
    tlMain = (TabLayout) findViewById(R.id.tlMain);
    tlMain.setTabMode(TabLayout.MODE_FIXED);
    tlMain.addTab(tlMain.newTab().setText(titleDatas.get(0)));
    tlMain.addTab(tlMain.newTab().setText(titleDatas.get(1)));
```

```java
        vpMain = (ViewPager) findViewById(R.id.vpMain);
        adapter = new MyAdapter(titleList, viewList);
        tlMain.setTabsFromPagerAdapter(adapter);
        vpMain.setAdapter(adapter);
        tlMain.setupWithViewPager(vpMain, true);
    }
```

(2) MyAdapter 实现。

```java
public class MyAdapter extends PagerAdapter {
    private ArrayList<String> titleList;
    private ArrayList<View> viewList;
    public MyAdapter(ArrayList<String> titleList, ArrayList<View> viewList) {
        this.titleList = titleList;
        this.viewList = viewList;
    }
    @Override
    public int getCount() {
        return viewList.size();
    }
    @Override
    public boolean isViewFromObject(View view, Object object) {
        return view == object;
    }
    @Override
    public Object instantiateItem(ViewGroup container, int position) {
        container.addView(viewList.get(position));
        return viewList.get(position);
    }
    @Override
    public void destroyItem(ViewGroup container, int position, Object object) {
        container.removeView(viewList.get(position));
    }
    @Override
    public CharSequence getPageTitle(int position) {
        return titleList.get(position);
    }
}
```

【拓展作业】

创建一个朋友圈软件，要求设计合理、美观，可参照应用商城应用模型。

单元八 Service 的生命周期

 课程目标

- Service 简介
- Service 生命周期
- 创建和使用 Service
- Service 的应用案例

 简 介

Service 通常被称为"后台服务",其中"后台"一词是相对于前台而言的,具体是指其本身的运行并不依赖于用户可视的用户界面。因此,从实际业务需求上来理解,Service 的适用场景应该具备以下条件:其一,并不依赖于用户可视的用户界面;其二,具有较长时间的运行特性。本单元将重点介绍 Service 的创建和使用。

8.1 Service 简介

Service(服务)是 Android 系统中的四大组件之一,是一种可以在后台执行长时间运行操作而没有用户界面的应用组件。服务可由其他应用组件(如 Activity)启动,服务一旦被启动将在后台一直运行,即使启动服务的组件已被销毁也不受影响。此外,组件可以绑定到服务,以与之进行交互,甚至是执行进程间的通信(IPC)。例如,服务可以处理网络事务、播放音乐、执行文件 I/O 或与内容提供程序交互等,这一切均可在后台进行。

服务分为两种,分别是本地服务(Local Service)和远程服务(Remote Service)。

1. 本地服务用于应用程序内部

本地服务用于应用程序内部。在 Service 可以调用 Context.startService()启动,调用 Context.stopService()结束。在内部可以调用 Service.stopSelf 或 Service.stopSelfResult()来自己停止。无论调用多少次 startService(),都只需调用一次 stopService()来停止。

2. 远程服务用于 Android 系统内部的应用程序之间

远程服务用于 Android 系统内部的应用程序之间。可以定义接口并把接口暴露出来,以便其他应用进行操作。客户端建立到服务对象的连接,并通过该连接来调用服务。调用 Context.bindService()方法建立连接并启动,以调用 Context.unbindService()关闭连接。多个客户端可以绑定至同一个服务。如果服务此时还没有加载,bindService()会先加载它,提供给可被其他应用复用,例如,定义一个天气预报服务,提供给其他应用调用即可。

8.2 Service 生命周期

Service 的生命周期方法比 Activity 要少,只有 onCreate()、onStart()、onDestroy()。启动 Service 有以下两种方式,它们对 Service 生命周期的影响是不一样的。

1. 启动状态

当应用组件(如 Activity)通过调用 startService()启动服务时，服务即处于"启动"状态。一旦启动，服务即可在后台无限期运行，即使启动服务的组件已被销毁也不受影响，除非手动调用才能停止服务，已启动的服务通常是执行单一操作，而且不会将结果返回给调用方。

2. 绑定状态

当应用组件通过调用 bindService()绑定到服务时，服务即处于"绑定"状态。绑定服务提供了一个客户端—服务器接口，允许组件与服务进行交互、发送请求、获取结果，甚至是利用进程间通信(IPC)跨进程执行这些操作。 仅当与另一个应用组件绑定时，绑定服务才会运行。多个组件可以同时绑定到该服务，但全部取消绑定后，该服务即会被销毁。

特别注意：

- 在调用 bindService()绑定到 Service 时，我们就应当保证在某处调用 unbindService()解除绑定(尽管 Activity 被停止的时候绑定会自动解除，而且 Service 会自动停止)。
- 使用 startService()启动服务之后，一定要使用 stopService()停止服务，不管我们是否使用 bindService()。
- 同时使用 startService()与 bindService()要注意，Service 的终止需要 unbindService()与 stopService()同时调用，不管 startService()与 bindService()的调用顺序，如果先调用 unbindService()，此时服务不会自动终止，再调用 stopService()之后服务才会停止，如果先调用 stopService()，此时服务也不会终止，只有再调用 unbindService()或者之前调用的 bindService()的 Context 不存在了(如 Activity 被 finish 的时候)之后服务才会自动停止。
- 当在旋转手机屏幕时，即手机屏幕在横、竖变换时，如果 Activity 会自动旋转，则是 Activity 的重新创建，因此旋转之前使用 bindService()建立的连接便会断开(即 Context 不存在了)，对应服务的生命周期与上述相同。
- 在 SDK 2.0 及其以后的版本中,对应的 onStart()已经被否决为了 onStartCommand()，但之前的 onStart()仍然有效。这意味着，如果我们开发的应用程序用的 SDK 为 2.0 及其以后的版本，那么我们应当使用 onStartCommand()而不是 onStart()。

图 8-1 所示的是用 startService()、bindService()两种方式启动 Service 的生命周期。

要使用服务，必须继承 Service 类(或者 Service 类的现有子类)，在子类中重写某些回调方法，以处理服务生命周期的某些关键方面并提供一种机制将组件绑定到服务。

1) onCreate()

首次创建服务时，系统将调用 onCreate()方法来执行初始化操作(在调用 onStartCommand()或onBind()之前)。如果在启动或绑定之前Service已在运行，则不会调用此方法。

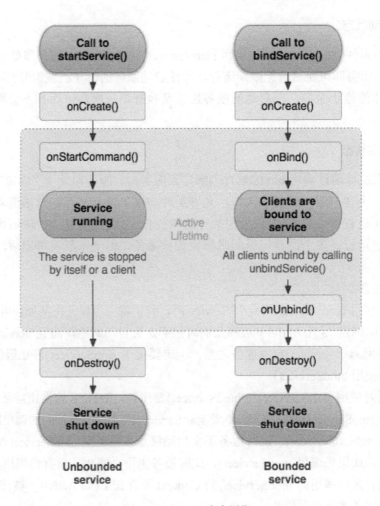

图 8-1 Service 生命周期

2) onStartCommand()

当组件通过调用 startService()请求启动服务时，系统将调用 onStartCommand()方法(如果是绑定服务，则不会调用此方法)。一旦执行此方法，服务即会启动并可在后台无限期运行。在指定任务完成后，通过调用 stopSelf()或 stopService()来停止服务。

3) onBind()

当一个组件想通过调用 bindService()与服务绑定时，系统将调用 onBind()方法(如果是启动服务，则不会调用此方法)。在 onBind()方法的实现中，必须通过返回 IBinder 提供一个接口，供客户端用来与服务进行通信。

4) onUnBind()

当另一个组件通过调用 unbindService()与服务解绑时，系统将调用 onUnBind()方法。只能调用 unbindService()一次，如重复多次调用此方法系统将会抛出错误异常。所以最简单的处理方式是设置一个静态变量 boolean connected，在调用 unbindService()前先做出判断。

5) onDestroy()

当服务不再使用且将被销毁时,系统将调用 onDestroy()方法,这是服务接收的最后一个调用,在此方法中应清理占用的资源。

仅当内存过低必须回收系统资源以供前台 Activity 使用时,系统才会强制停止服务。如果将服务绑定到前台 Activity,则它不太可能会终止;如果将服务声明为在前台运行,则它几乎永远不会终止。或者,如果服务已启动并要长时间运行,则系统会随着时间的推移降低服务在后台任务列表中的位置,而服务也将随之变得非常容易被终止。如果服务是启动服务,则必须将其设计为能够妥善处理系统对它的重启。如果系统终止服务,那么一旦资源变得再次可用,系统便会重启服务。

8.2.1 创建 Service

创建 Service 的操作步骤如下。

(1) 继承 Service 类,实现自己的 Service。

如果想要访问 Service 中的某些值,通常会提供一个继承了 Binder 的内部类,通过 onBund()方法返回给 Service 请求。这里实际上巧妙地利用了内部类能够访问外部类属性的特点。

(2) 在 androidManifest.xml 中进行注册,代码如下。

```
<!-- Service 配置开始 -->
 <service android:name="MyService"></service>
<!-- Service 配置结束-->
```

(3) 在其他组件中进行启动、绑定、解绑或者停止 Service 操作。

8.2.2 startService()启动服务

通过 startService()启动的服务,Service 的生命周期过程为 startService()→onCreate()→onStartCommand()→Service is running→stopService()→onDestroy()。根据上述创建服务的步骤,首先自定义一个服务类 SimpleService,代码如下。

```
public class SimpleService extends Service {
public IBinder onBind(Intent intent) {return null;}
    public void onCreate() {
        super.onCreate();
        System.out.println("Service is created!!!");
        Toast.makeText(Service1.this, "service created....", Toast.LENGTH_LONG).show();
    }
    public void onDestroy() {
        super.onDestroy();
        System.out.println("Service is destroyed!!!");
        Toast.makeText(Service1.this, "service destroyed....", Toast.LENGTH_LONG).show();
    }
```

```
public int onStartCommand(Intent intent, int flags, int startId) {
    System.out.println("Service is started!!!");
    Toast.makeText(Service1.this, "service started....", Toast.LENGTH_LONG).show();
    return START_STICKY;}}
```

从上面的代码中我们可以看出SimpleService继承了Service类，并重写了onBind()方法。该方法必须重写，但是由于此时是启动状态的服务，所以该方法无须实现，返回null即可，只有在绑定状态的情况下才需要实现该方法并返回一个IBinder的实现类(将在后面详细讲解)，接着重写了onCreate()、onStartCommand()、onDestroy() 3个主要的生命周期方法。

自定义Service后，务必在清单配置文件中声明Service，声明方式与Activity相似，代码如下。

```xml
<manifest ... >
  ...
  <application ... >

<service    android:name=".service.SimpleService " >
    <intent-filter >
        <action android:name="hp.service.SimpleService " />
    </intent-filter>
 </service>

    ...
  </application>
</manifest>
```

MainActivity.java：

```java
import android.content.Intent;
import android.os.Bundle;
import android.support.annotation.Nullable;
import android.support.v7.app.AppCompatActivity;
import android.view.View;
import android.widget.Button;

import com.hp.main.R;

public class MainActivity extends AppCompatActivity{
    private Button start;
    private Button stop;
    @Override
    protected void onCreate(@Nullable Bundle savedInstanceState) {
        super.onCreate(savedInstanceState);
        setContentView(R.layout.activity_main);
        start = (Button)findViewById(R.id.start);
        stop = (Button)findViewById(R.id.stop);
        start.setOnClickListener(new View.OnClickListener() {
            @Override
```

```
                public void onClick(View v) {
                    Intent intent = new Intent(MainActivity.this,SimpleService.class);
                    startService(intent);
                }
            });
            stop.setOnClickListener(new View.OnClickListener() {
                @Override
                public void onClick(View v) {
                    Intent intent = new Intent(MainActivity.this,SimpleService.class);
                    stopService(intent);
                }
            });
        }
    }
```

从以上代码可以看到，在 start 按钮的单击事件里，我们构建出了一个 Intent 对象，并调用 startService()方法来启动 SimpleService。在 stop 按钮的单击事件里，我们同样构建出了一个 Intent 对象，并调用 stopService()方法来停止 SimpleService。代码的逻辑非常简单，这里不再赘述。

至此，一个简单的带有 Service 功能的程序就写好了，现在我们运行程序，并单击 start 按钮，可以看到 LogCat 的打印日志如图 8-2 所示。

```
D/SimpleService: Service is created!!!
D/SimpleService: Service is started!!!
```

图 8-2

也就是说，当启动一个 Service 时，会调用该 Service 中的 onCreate()和 onStartCommand()方法。

我们再单击 stop 按钮，打印日志如图 8-3 所示。

```
D/SimpleService: Service is destroyed!!!
```

图 8-3

可以看到，只有 onStartCommand()方法执行了，而 onCreate()方法并没有执行，为什么会这样呢？这是由于 onCreate()方法只会在 Service 第一次被创建的时候调用，如果当前 Service 已经被创建过了，不管怎样调用 startService()方法，onCreate()方法都不会再执行。我们可以多单击几次 Start Service 按钮试一试，结果是每次都只会有 onStartCommand()方法中的打印日志。

我们还可以在手机的应用程序管理界面检查 SimpleService 是否正在运行，如图 8-4 和图 8-5 所示。

图 8-4　　　　　　　　　　　　　　图 8-5

8.2.3　bindService()绑定服务

通过 bindService()启动服务的过程为：context.bindService()→onCreate()→onBind()→Service running→onUnbind()→onDestroy()。

修改 MyService 中的代码，如下所示。

```
import android.app.Service;
import android.content.Intent;
import android.os.Binder;
import android.os.IBinder;
import android.util.Log;

public class MyService extends Service {

    public static final String TAG = "MyService";

    private MyBinder mBinder = new MyBinder();

    @Override
    public void onCreate() {
        super.onCreate();
```

```java
        Log.d(TAG, "onCreate() executed");
    }

    @Override
    public int onStartCommand(Intent intent, int flags, int startId) {
        Log.d(TAG, "onStartCommand() executed");
        return super.onStartCommand(intent, flags, startId);
    }

    @Override
    public void onDestroy() {
        super.onDestroy();
        Log.d(TAG, "onDestroy() executed");
    }

    @Override
    public IBinder onBind(Intent intent) {
        return mBinder;
    }

    class MyBinder extends Binder {

        public void startDownload() {
            Log.d("TAG", "startDownload() executed");
            // 执行具体的下载任务
        }
    }
}
```

这里我们新增了一个 MyBinder 类，继承自 Binder 类，然后在 MyBinder 中添加一个 startDownload()方法用于在后台执行下载任务，由于这里并不是真正地去下载某个东西，只是做个测试，所以 startDownload()方法只是打印了一行日志。

修改 activity_main.xml 中的代码，在布局文件中添加用于绑定 Service 和取消绑定 Service 的按钮。

```java
import android.content.ComponentName;
import android.content.Intent;
import android.content.ServiceConnection;
import android.os.Bundle;
import android.os.IBinder;
import android.support.annotation.Nullable;
import android.support.v7.app.AppCompatActivity;
import android.view.View;
import android.widget.Button;

import com.hp.main.R;
```

```java
public class MainActivity extends AppCompatActivity{
    private Button bind;
    private Button unbind;
    private MyService.MyBinder myBinder;
    private ServiceConnection connection = new ServiceConnection() {

        @Override
        public void onServiceDisconnected(ComponentName name) {

        }

        @Override
        public void onServiceConnected(ComponentName name, IBinder service) {
            myBinder = (MyService.MyBinder) service;
            myBinder.startDownload();
        }
    };
    @Override
    protected void onCreate(@Nullable Bundle savedInstanceState) {
        super.onCreate(savedInstanceState);
        setContentView(R.layout.activity_main);
        bind = (Button)findViewById(R.id.bind);
        unbind = (Button)findViewById(R.id.unbind);
        bind.setOnClickListener(new View.OnClickListener() {
            @Override
            public void onClick(View v) {
                Intent bindIntent = new Intent(MainActivity.this,MyService.class);
                bindService(bindIntent,connection,BIND_AUTO_CREATE);
            }
        });
        unbind.setOnClickListener(new View.OnClickListener() {
            @Override
            public void onClick(View v) {
                unbindService(connection);
            }
        });
    }
}
```

这里我们首先创建了一个 ServiceConnection 的匿名类，并在里面重写了 onServiceConnected()方法和 onServiceDisconnected()方法，这两个方法分别在 Activity 与 Service 建立关联和解除关联的时候调用。在 onServiceConnected()方法中，我们可以通过向下转型得到 MyBinder 的实例，这样使 Activity 和 Service 之间建立了联系。现在我们可以在 Activity 中根据具体的场景来调用 MyBinder 中的任何 public()方法，即实现了 Activity 可指挥 Service 做任意事情的功能，然后在 bind 按钮的单击事件中完成。可以看到，这里我们仍然是构建出了一个 Intent 对象,然后调用 bindService()方法将 Activity 和 Service 进行绑定。bindService()方法接收三个参数:第一个参数是刚刚构建出的 Intent 对象;第二个参数是前面创建出的 ServiceConnection 的实例;第三个参数是一个标志位,这里传入 BIND_AUTO_CREATE 表示在 Activity 和 Service 建立关联后自动创建 Service,这会使得 MyService 中的 onCreate()方法得到执行,但 onStartCommand()方法不会执行。如果我们想解除 Activity 和 Service 之间的关联,则调用 unbindService()方法即可,这也是 Unbind Service 按钮的单击事件里实现的逻辑。

现在我们重新运行一下程序,在 MainActivity 中单击 Bind Service 按钮,LogCat 中的打印日志如图 8-6 所示。

```
D/MyService: onCreate() executed
D/TAG: startDownload() executed
D/MyService: onDestroy() executed
```

图 8-6

现在我们可以在手机的应用程序管理界面检查一下 MyService 是不是正在运行,如图 8-7 和图 8-8 所示。

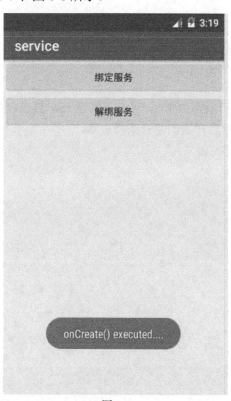

图 8-7

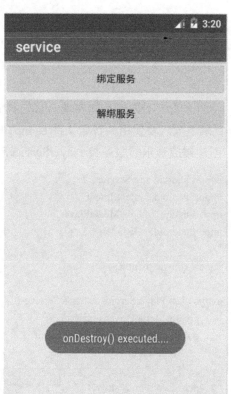

图 8-8

8.3 简易音乐播放器

简易音乐播放器的界面如图 8-9 所示

图 8-9

下面来看一下如何具体实现，首先我们在应用程序的 res 目录下创建 raw 目录，并在 raw 目录下放入要使用的音乐文件 aa.mp3 中，然后开始创建音乐播放器。

创建播放音乐的服务类 PlayerService.java 代码如下：

```
import android.app.Service;
import android.content.Intent;
import android.media.MediaPlayer;
import android.os.IBinder;

import com.hp.main.R;

public class PlayerService extends Service {
    private MediaPlayer player;
    public IBinder onBind(Intent intent) {
        return null;
    }
    public void onDestroy() {
        super.onDestroy();
        player.stop();
    }

    @Override
```

```
        public void onCreate() {
            super.onCreate();
            player=MediaPlayer.create(this, R.raw.aa);
            player.start();
        }

}
```

从上面的代码中我们可以看出 PlayerService 继承了 Service 类，在 onCreate()方法中我们通过 create 实例化 MediaPlayer，并通过调用 MediaPlayer 实例的 start()方法开始音乐的播放。

修改 MainActivity.java 代码如下。

```
import android.app.Activity;
import android.content.Intent;
import android.os.Bundle;
import android.view.View;
import android.view.View.OnClickListener;
import android.widget.Button;

import com.hp.main.R;

public class MainActivity extends Activity {
    private   Button start,stop;
    public void onCreate(Bundle savedInstanceState) {
        super.onCreate(savedInstanceState);
        setContentView(R.layout.activity_main);
        start=(Button) findViewById(R.id.start);
        stop=(Button) findViewById(R.id.stop);
        final Intent intent=new Intent(this,PlayerService.class);
        start.setOnClickListener(new OnClickListener(){
            public void onClick(View v) {
            startService(intent);
            }
        });
        stop.setOnClickListener(new OnClickListener(){
            public void onClick(View v) {
                stopService(intent);
            }
        });
    }
}
```

从以上代码中可以看到，在 start 按钮的单击事件中，我们开启服务，就开始播放音乐；在 stop 按钮的单击事件里，我们停止服务，就停止播放音乐。代码的逻辑非常简单，这里不再赘述。

当然，与创建自己的 Activity 一样，创建自己的服务同样需要在 androidManifest.xml 文件中对自定义的 Service 组件进行注册，代码如下：

```
<!-- Service 配置开始-->
<service android:name="com.hp.music.PlayerService">
<!-- Service 配置结束-->
```

【单元小结】

- Service 概述。
- Service 简介。
- 远程调用 Service。

【单元自测】

1. 下列 Android 关于 Service 生命周期的 onCreate()和 onStart()说法，正确的是(　　)。
   A. 当第一次启动的时候先后调用 onCreate()和 onStart()方法
   B. 当第一次启动的时候只会调用 onCreate()方法
   C. 如果 Service 已经启动，将先后调用 onCreate()和 onStart()方法
   D. 如果 Service 已经启动，只会执行 onStart()方法，不再执行 onCreate()方法
2. Android 中 Service 实现的方法有(　　)。
   A. startService()   B. beginService()
   C. bindService()   D. initService()
3. Service 的生命周期的方法有(　　)。
   A. onCreate()   B. onStart()
   C. onDestroy()   D. onStop()
4. Android 绑定 Service 的方法是(　　)。
   A. bindService()   B. startService()
   C. onStart()   D. onBind()
5. 使用 Service 的步骤依次为(　　)。
   A. 在 androidManifest.xml 中进行注册
   B. 我们要继承 Service 类，实现自己的 Service
   C. 在 Activity 中启动、绑定、解绑或者停止 Service
   D. 我们要继承 AndroidService 类，实现自己的 Service

【上机实战】

上机目标

- 了解 Service 的生命周期。
- 在 startService 和 bindService 时，Service 如何响应。

上机练习

练习：掌握 Service 生命周期的特性

【问题描述】
如何绑定服务和启动服务。

【问题分析】
(1) 编写 Android Service 需要基础 Service 类，并实现其中的 onBind()方法。
(2) 在 AndroidManifest.xml 文件中声明 Service 组件。
(3) 在需要 Service 的地方通过 Context.startService(Intent)方法启动 Service 或者通过 Context.bindService()方法绑定 Service。

【参考步骤】
(1) 创建 Android 工程，效果如图 8-10 所示。

图 8-10

代码如下。

```
/**
 * Android Service 示例
 *
 * @author dev
 *
```

```java
    */
    public class ServiceDemo extends Service {
        private static final String TAG = "ServiceDemo" ;
        public static final String ACTION = "com.lql.service.Servic
            eDemo";
        @Override
        public IBinder onBind(Intent intent) {
            Log.v(TAG, "ServiceDemo onBind");
            return null;
        }

        @Override
        public void onCreate() {
            Log.v(TAG, "ServiceDemo onCreate");
            super.onCreate();
        }

        @Override
        public void onStart(Intent intent, int startId) {
            Log.v(TAG, "ServiceDemo onStart");
            super.onStart(intent, startId);
        }

        @Override
        public int onStartCommand(Intent intent, int flags, int startId) {
            Log.v(TAG, "ServiceDemo onStartCommand");
            return super.onStartCommand(intent, flags, startId);
        }
    }
```

(2) 在 AndroidManifest.xml 文件中声明 Service 组件。

XML 代码如下。其中，intent-filter 中定义的 action 是用来启动服务的 Intent。

```xml
<service android:name="com.lql.service.ServiceDemo">
    <intent-filter>
        <action android:name="com.lql.service.ServiceDemo"/>
    </intent-filter>
</service>
```

(3) 启动 Service 或绑定 Service。

Java 代码如下。

```java
public class ServiceDemoActivity extends Activity {
    private static final String TAG = "ServiceDemoActivity";

    Button bindBtn;
    Button startBtn;
    @Override
    public void onCreate(Bundle savedInstanceState) {
```

```java
        super.onCreate(savedInstanceState);
        setContentView(R.layout.main);

        bindBtn = (Button)findViewById(R.id.bindBtn);
        startBtn = (Button)findViewById(R.id.startBtn);

        bindBtn.setOnClickListener(new OnClickListener() {
            public void onClick(View v) {
                bindService(new Intent(ServiceDemo.ACTION), conn, BIND_AUTO_CREATE);
            }
        });

        startBtn.setOnClickListener(new OnClickListener() {
            public void onClick(View v) {
                startService(new Intent(ServiceDemo.ACTION));
            }
        });
    }
    ServiceConnection conn = new ServiceConnection() {
        public void onServiceConnected(ComponentName name, IBinder service) {
            Log.v(TAG, "onServiceConnected");
        }
        public void onServiceDisconnected(ComponentName name) {
            Log.v(TAG, "onServiceDisconnected");
        }
    };
    @Override
    protected void onDestroy() {
        Log.v(TAG, "onDestroy unbindService");
        unbindService(conn);
        super.onDestroy();
    };
}
```

(4) 日志输出，结果如图 8-11 所示。

```
D/ServiceDemo: ServiceDemo    onCreate
D/ServiceDemo: ServiceDemo    onBind
```

图 8-11

图 8-11 是单击绑定服务时输出的。可以看出，只调用了 onCreate()方法和 onBind()方法，当重复单击绑定服务时，没有再输出任何日志，并且不报错。onCreate()方法是在第一次创建 Service 时调用的，而且只调用一次。另外，在绑定服务时，给定了参数 BIND_AUTO_CREATE，即当服务不存在时，自动创建，如果服务已经启动或创建，那么只会调用 onBind()方法，如图 8-12 所示。

```
D/ServiceDemo: ServiceDemo   onCreate
D/ServiceDemo: ServiceDemo   onStartCommand
D/ServiceDemo: ServiceDemo   onStart
D/ServiceDemo: ServiceDemo   onStartCommand
D/ServiceDemo: ServiceDemo   onStart
D/ServiceDemo: ServiceDemo   onStartCommand
D/ServiceDemo: ServiceDemo   onStart
```

图 8-12

图 8-12 是在多次单击启动服务时输出的。可以看出，在第一次单击时，因为 Service 还未创建，所以调用了 onCreate()方法，接着调用了 onStartCommand()和 onStart()方法。当再次单击启动服务时，仍然调用了 onStartCommand()和 onStart()方法，所以，在 Service 中做任务处理时需要注意这一点，因为一个 Service 可以被重复启动。

这里说明一下，平常使用较多的是 startService()方法，可以把一些耗时的任务放到后台处理，当处理完成后，可以通过广播通知前台。而 onBind()方法更多的是结合 AIDL 使用，这样一个应用可以通过绑定服务获得的 IBinder 拿到后台的接口，进而调用 AIDL 中定义的方法，进行数据交换等。

【拓展作业】

1. 参考上机部分编写一个实例，测试 Service 生命周期。
2. 在模拟器和控制台上分析、观察 Service 生命周期的变化。

单元九 广播和通知

 课程目标

- ▶ 自定义 BroadcastReceiver
- ▶ 系统广播事件使用
- ▶ 通知

 简介

Android 广播分为两个方面：广播发送者和广播接收者，通常情况下，BroadcastReceiver 指的就是广播接收者(广播接收器)。广播作为 Android 组件间的通信方式，可以使用的场景如下：同一 APP 内部的同一组件内的消息通信(单个或多个线程之间)；同一 APP 内部的不同组件之间的消息通信(单个进程)；同一 APP 具有多个进程的不同组件之间的消息通信；不同 APP 组件之间的消息通信；Android 系统在特定情况下与 APP 之间的消息通信。从实现原理上看，Android 中的广播使用了观察者模式，基于消息的发布/订阅事件模型。因此，从实现的角度来看，Android 中的广播将广播的发送者和接收者的极大程度的解耦，使得系统能够方便集成，更易扩展。本单元将会重点介绍广播机制。

9.1 广播机制简介

为什么说 Android 中的广播机制更加灵活呢？这是因为 Android 中的每个应用程序都可以对自己感兴趣的广播进行注册，这样该程序就只会接收到自己所关心的广播内容，这些广播可能来自于系统，也可能来自于其他应用程序。Android 提供了一套完整的 API，允许应用程序自由地发送和接收广播，如图 9-1 所示。

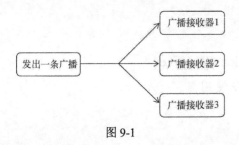

图 9-1

发送广播的方法需要借助单元二中学过的 Intent，而接收广播的方法则需要引入一个新的概念——广播接收器(BroadcastReceiver)。广播接收器的具体用法将会在下一节中做介绍，这里我们先来了解一下广播的类型。Android 中的广播主要可以分为两种类型：标准广播和有序广播。

有序广播(Ordered Broadcasts)是一种同步执行的广播，在广播发出之后，同一时刻只会有一个广播接收器能够收到这条广播消息，当这个广播接收器中的逻辑执行完毕后，广播才会继续传递。所以此时的广播接收器是有先后顺序的，优先级高的广播接收器可以先收到广播消息，并且前面的广播接收器还可以截断正在传递的广播，这样后面的广播接收器就无法收到广播消息了。有序广播的工作流程如图 9-2 所示。

掌握了这些基本概念后，我们就可以来尝试一下广播的用法了，首先从接收系统广播开始。

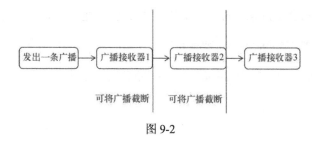

图 9-2

9.2 系统广播事件使用

如果应用需要在系统特定时刻执行某些操作，则可以通过监听系统广播来实现。Android 的很多系统事件都会对外发送标准的系统广播。以下是常见的 Android 广播 Action 常量(详细内容请参考 Android API 文档中关于 Intent 的详细用法)。

- ACTION_TIME_CHANGED：系统时间被改变。
- ACTION_DATE_CHANGED：系统日期被改变。
- ACTION_TIMEZONE_CHANGED：系统时区被改变。
- ACTION_BOOT_COMPLETED：系统启动完成。
- ACTION_PACKAGE_ADDED：系统中添加安装包。
- ACTION_PACKAGE_CHANGED：系统的包被改变。
- ACTION_PACKAGE_REMOVED：系统的包被删除。
- ACTION_PACKAGE_RESTARTED：系统的包被重启。
- ACTION_PACKAGE_DATA_CLEARED：系统的包数据被清空。
- ACTION_BATTERY_CHANGED：电池电量改变。
- ACTION_SHUTDOWN：系统被关闭。
- ACTION_BATTRY_LOW：电池电量低。

通过监听特殊的广播，即可实现应用跟随系统执行特定的某些操作。下面通过示例处理系统事件，新建一个类。

MyReceiverSystem.java：

```java
package com.ex;
import android.content.BroadcastReceiver;
import android.content.Context;
import android.content.Intent;
import android.util.Log;
public class MyReceiverSystem  extends BroadcastReceiver{
    @Override
    public void onReceive(Context context, Intent intent) {
        // TODO Auto-generated method stub
        Log.i("my_tag","BOOT_COMPLETED~~~~~~~~~~~~~~~~");
    }
}
```

此外，在 AndroidManifest.xml 中也要对这个类进行注册，添加的代码如下。

```xml
<receiver android:name=".MyReceiverSystem">
    <intent-filter>
        <action android:name="android.intent.action.BOOT_COMPLETED"/>
    </intent-filter>
</receiver>
```

上面的案例中，我们是通过AndroidManifest注册广播接收器的，当然也可以通过代码的方式来注册或注销一个广播接收器：使用Activity.onResume()方法中的Context.registerReceiver()方法注册；使用Activity.onPause()方法中的Context.unregisterReceiver(r)方法注销。代码如下。

```
//实例化 IntentFilter
IntentFilter filter=new IntentFilter();
//实例化 Receiver
MyReceiverSystem r = new MyReceiverSystem();
//注册 Receiver
registerReceiver(r,filter);
//注销 Receiver
unregisterReceiver(r);
```

9.3 自定义 BroadcastReceiver

BroadcastReceiver 处理的是系统级别的监听，事件的广播机制是构建 Intent 对象，再调用 sendBroadcast()方法将广播发出去，事件的接收则是通过继承一个 BroadcastReceiver 的类来实现，覆盖其 onReceive()方法。Android 系统中定义了很多标准的 Broadcast Action 用来响应系统广播事件，如 ACTION_TIME_CHANGED(时间改变时触发)、ACTION_BOOT_COMPLETED(系统启动完成后触发)、ACTION_PACKAGE_ADDED(添加安装包时触发)、ACTION_BATTERY_CHANGED(电池电量改变时触发)，当然也可以自定义 Action。

main.xml：

```xml
<?xml version="1.0" encoding="utf-8"?>
<LinearLayout xmlns:android="http://schemas.android.com/apk/res/android"
    android:orientation="vertical"
    android:layout_width="fill_parent"
    android:layout_height="fill_parent"
    >
<Button
    android:text="SEND BROADCAST"
    android:id="@+id/btn_send"
    android:layout_width="wrap_content"
    android:layout_height="wrap_content"
    />
</LinearLayout>
```

MainActivity.java：

```java
package com.ex;

import android.app.Activity;
import android.content.Intent;
import android.os.Bundle;
import android.view.View;
import android.view.View.OnClickListener;
import android.widget.Button;

public class MainActivity extends Activity {
    /** Called when the activity is first created. */

    private static final String MY_ACTION ="com.ex.action.MY_ACTION";
    private Button btn;

    @Override
    public void onCreate(Bundle savedInstanceState) {
        super.onCreate(savedInstanceState);
        setContentView(R.layout.main);
        btn=(Button)findViewById(R.id.btn_send);
        btn.setOnClickListener(new OnClickListener(){

            @Override
            public void onClick(View v) {
                // TODO Auto-generated method stub
                Intent intent = new Intent();//实例化 intent 对象
                intent.setAction(MY_ACTION);//设置 Intent Action 属性
                //为 Intent 添加附加信息
                intent.putExtra("msg", "received in MyBroadcastReceive");
                //发布广播
                sendBroadcast(intent);
            }
        });
    }
}
```

MyReceiver.java：

```java
package com.ex;
//接收广播
import android.content.BroadcastReceiver;
import android.content.Context;
import android.content.Intent;
import android.widget.Toast;

public class MyReceiver extends BroadcastReceiver {
```

```java
        @Override
        public void onReceive(Context context, Intent intent) {
            // TODO Auto-generated method stub
            //从 Intent 中获得信息
            String msg=intent.getStringExtra("msg");
            //使用 Toast 显示
            Toast.makeText(context, msg, Toast.LENGTH_LONG).show();
        }
    }
```

AndroidManifest.xml:

```xml
<receiver android:name=".MyReceiver"
    android:label="@string/app_name">
        <intent-filter>
        <action android:name=
            "com.ex.action.MY_ACTION"/>
            </intent-filter>
        </receiver>
```

效果如图 9-3 所示。

总结：

- 发出广播需要一个发出广播的类(当然也可以有多个)，构建 intent 对象，然后调用 sendBroadcast(intent 对象)。
- 需要一个接收广播的类，并且这个类继承 BroadcactReceiver，要先接收 Intent 中的信息。
- 要在AndroidManifest中进行Receiver的注册。

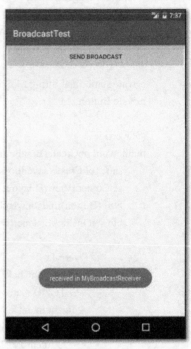

图 9-3

9.4 通知

通知即 Notifaction。使用 Notifaction 的目的是，BroadcastReceiver 组件并没有提供可视化界面来显示广播信息，而 Notifaction 可以实现可视化的信息显示，通过它可以显示广播信息的内容及图标和震动等信息(在状态栏上)。

通知是 Android 系统中比较有特色的一个功能，当应用程序希望向用户发出提示信息，而该应用程序又不在前台运行时，就可以借助通知来实现。发出一条通知后，手机最上方的状态栏中会显示一个通知的图标，下拉状态栏后可以看到通知的详细内容。

使用 Notifaction 显示通知的具体步骤如下。

(1) 获得系统级的服务 Notification Manager 通知管理，调用 getSystemService()方法实现。

```
String service = NOTIFICATION_SERVICE;
nm=(NotificationManager)getSystemService(service);
```

(2) 实例化 Notification，设置其属性(要显示的)。

```
//实例化 Notification
n = new Notification();
//设置显示图标，该图标会在状态栏显示
int icon = n.icon=R.drawable.notification;
//设置显示提示信息，该信息会在状态栏显示
String tickerText="Test Notification";
//显示时间
long when = System.currentTimeMillis();
n.icon=icon;
n.tickerText=tickerText;
n.when=when;
```

当然也可以通过下面的代码实现。

```
Notification n1 = new Notification(icon,tickerText,when);
```

此外，设置视图中的图标和时间可以通过以下代码实现。

```
//实例化 intent
Intent intent = new Intent(MainActivity.this,MainActivity.class);
//获取 PendingIntent，PendingIntent 是单击通知后所跳转的页面
PendingIntent pi = PendingIntent.getActivity(MainActivity.this, 0, intent, 0);
////新建 Notification.Builder 对象设置事件信息
Notification.Builder builder = new Notification.Builder(MainActivity.this);
            builder.setContentTitle("Bmob Test");
            builder.setContentText("message");
            builder.setSmallIcon(R.drawable.notification);
            builder.setContentIntent(pi);//执行 intent
            n=builder.getNotification();//将 builder 对象转换为普通的 notification
            n.flags=Notification.FLAG_AUTO_CANCEL;//单击通知后通知消失
            nm.notify(1,n);//运行 notification//发出通知
```

完整的代码如下。

```
package com.example.luyong.myapplication;

import android.app.Notification;
import android.app.NotificationManager;
import android.app.PendingIntent;
import android.content.Intent;
import android.support.v7.app.AppCompatActivity;
import android.os.Bundle;
```

```java
import android.view.View;
import android.widget.Button;
import android.view.View.OnClickListener;

public class MainActivity extends AppCompatActivity {
    Button btn_send,btn_cancel;
    //声明 Notifaction
    private Notification n;
    //声明 NotificationManager
    private NotificationManager nm;
    //Notification 标识 ID
    private static final int ID = 1;

    @Override
    protected void onCreate(Bundle savedInstanceState) {
        super.onCreate(savedInstanceState);
        //实例化两个按钮
        setContentView(R.layout.activity_main);
        btn_send=(Button)findViewById(R.id.btn_send);
        btn_cancel=(Button)findViewById(R.id.btn_cancel);
        //获得 NotifactionManager 实例
        String service = NOTIFICATION_SERVICE;
        nm=(NotificationManager)getSystemService(service);
        //设置显示提示信息，该信息会在状态栏显示
        String tickerText="Test Notification";
        //显示时间
        long when = System.currentTimeMillis();
        btn_send.setOnClickListener(new OnClickListener() {
            @Override
            public void onClick(View view) {

            }
        });
        btn_send.setOnClickListener(new OnClickListener() {
            @Override
            public void onClick(View view) {
                //实例化 intent
                Intent intent = new Intent(MainActivity.this,MainActivity.class);
                //获取 pendingIntent
                PendingIntent pi = PendingIntent.getActivity(MainActivity.this, 0, intent, 0);
                //设置事件信息
                Notification.Builder builder = new Notification.Builder(MainActivity.this);
                                                        //新建 Notification.Builder 对象
                //PendingIntent 单击通知后所跳转的页面
                builder.setContentTitle("Bmob Test");
                builder.setContentText("message");
                builder.setSmallIcon(R.drawable.notification);
```

```
                builder.setContentIntent(pi);//执行 intent
                n=builder.getNotification();//将 builder 对象转换为普通的 notification
                n.flags=Notification.FLAG_AUTO_CANCEL;//单击通知后通知消失
                nm.notify(1,n);//运行 notification
                //发出通知
            }
        });

        btn_cancel.setOnClickListener(new View.OnClickListener(){

            @Override
            public void onClick(View view) {
                nm.cancel(ID);

            }});

    }
}
```

main.xml：

```
<?xml version="1.0" encoding="utf-8"?>
<LinearLayout xmlns:android="http://schemas.android.com/apk/res/android"
    android:orientation="vertical"
    android:layout_width="fill_parent"
    android:layout_height="fill_parent"
    >
<TextView
    android:layout_width="fill_parent"
    android:layout_height="wrap_content"
    android:text="测试 Notifaction"
    />
 <Button
     android:id="@+id/btn_send"
     android:text="发出通知"
     android:layout_width="wrap_content"
     android:layout_height="wrap_content"
 ></Button>
 <Button
     android:id="@+id/btn_cancel"
     android:text="取消通知"
     android:layout_width="wrap_content"
     android:layout_height="wrap_content"
  />
</LinearLayout>
```

效果如图 9-4 所示。

单击"发出通知"按钮后状态栏的显示如图 9-5 所示。

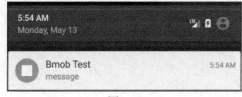

图 9-4   图 9-5

(3) 使用 NotificationManager 取消通知。

取消通知的两种方式上面代码已经全部展示过，在这里我们总结一下。

① 单击后自动取消：.setAutoCancel(true)//单击后，自动消失。

n.flags=Notification.FLAG_AUTO_CANCEL;//单击通知后通知消失

② 根据 id 取消：使用 NotificationManager 的 cancel(ID);。

```
//实例化 NotificationManager
    NotificationManager  nm=
        (NotificationManager) getSystemService(Context.NOTIFICATION_SERVICE);
    mNotificationManager .cancel(ID);
```

【单元小结】

- 自定义 BroadcastReceiver。
- 系统广播事件使用。
- 通知的使用。

【单元自测】

1. 下面在 AndroidManifest.xml 文件中注册 BroadcastReceiver 方式正确的是(   )。

A.
```
<receiver android:name="NewBroad">
    <intent-filter>
            <action  android:name="android.provider.action.NewBroad"/>
     <action>
    </intent-filter>
```

```
    </receiver>
```

B.
```
<receiver android:name="NewBroad">
  <intent-filter>
    android:name="android.provider.action.NewBroad"/>
  </intent-filter>
</receiver>
```

C.
```
<receiver android:name="NewBroad">
 <action
    android:name="android.provider.action.NewBroad"/>
 <action>
</receiver>
```

D.
```
<intent-filter>
<receiver android:name="NewBroad">
        <action>
          android:name="android.provider.action.NewBroad"/>
        <action>
</receiver>
</intent-filter>
```

2. BroadCastReceiver 实现的是(　　)。

　　A. 广播　　　　　　　　　　B. 接收

　　C. 广播的接收　　　　　　　D. 广播和广播的接收

3. 在(　　)方法中使用(　　)方法来注册一个广播接收器。

　　A. Activity.onResume()　　　　B. Context.registerReceiver()

　　C. Activity.Resume()　　　　　D. Context.OnRegisterReceiver()

4. 在(　　)方法中使用(　　)方法来注销一个广播接收器。

　　A. Activity.Pause()　　　　　　B. Context.UnregisterReceiver(r)

　　C. Activity.onPause()　　　　　D. Context.unregisterReceiver(r)

【上机实战】

上机目标

- 理解如何使用 BroadcastReceiver 进行广播的注册。
- 理解如何使用 BroadcastReceiver 进行广播的注销。

上机练习

【问题描述】
如何使用广播接收器进行广播的注册与注销。

【问题分析】
对广播接收器、注册广播、注销广播、在线通知等各个环节的特性进行分析。

【参考步骤】
(1) 新建广播接收器。

```java
package com.hp.broadcastreceiver;
import com.g4vision.R;
import com.g4vision.common.CONSTANT;
import com.g4vision.common.Utils;
import android.app.Notification;
import android.app.PendingIntent;
import android.content.BroadcastReceiver;
import android.content.Context;
import android.content.Intent;
import android.os.Bundle;

/**
 * 广播接收器
 *
 *
 */
public class MVRBroadcastReceiver extends BroadcastReceiver{
    private Context context;
    private Intent intent;
    @Override
    public void onReceive(Context context, Intent intent) {
        // TODO Auto-generated method stub
        this.context=context;
        this.intent=intent;
        Bundle bundle=intent.getExtras();
        String tmpInlineStatus=bundle.getString(CONSTANT.MVR_BROADCAST_CMD);
        //System.out.println("接收到的消息为"+tmpInlineStatus);
        if(tmpInlineStatus.equalsIgnoreCase("inline")){
            setInlineStatus();
        }else if(tmpInlineStatus.equalsIgnoreCase("uninline")){
            Utils.mNM.cancel(CONSTANT.INLINE_STATUS_ID);
        }else if(tmpInlineStatus.equalsIgnoreCase("picTransmitting")){
            setPicTransmittingStatus();
        }
    }
```

```java
/**
 * 设置在线图标
 */
private void setInlineStatus()
{
    Notification m=new Notification();
    m.icon=R.drawable.notification_inline_status;
    m.tickerText="inline";
    m.when=System.currentTimeMillis();
    //m.defaults |= Notification.DEFAULT_SOUND;
    // m.vibrate = new long[] { 100, 250, 100, 500};
    PendingIntent p=PendingIntent.getActivity(context, 0, intent, PendingIntent.FLAG_UPDATE_CURRENT);
    m.setLatestEventInfo(context, "Inline", "I am inline", p);
    Utils.mNM.notify(CONSTANT.INLINE_STATUS_ID, m);
}

/**
 * 指示图片正在传输
 */
private void setPicTransmittingStatus()
{
    //初始化 NotificationManager 对象
    Notification mNotification=new Notification();//实例化 mNotification
        //设置传输状态的图标
    mNotification.icon=R.anim.stat_transmit_upload;
    mNotification.tickerText="Pic Transmit";
    mNotification.when=System.currentTimeMillis();
    //通知到来时发出默认的声音，即震动
    //mNotification.defaults |= Notification.DEFAULT_SOUND;
    //mNotification.vibrate = new long[] { 100, 250, 100, 500};
    PendingIntent transmitPendingIntent=PendingIntent.getActivity(context, 0, intent,
        PendingIntent.FLAG_UPDATE_CURRENT);
    mNotification.setLatestEventInfo(context, "Transmit", "pic is transmitting",
        transmitPendingIntent);
    Utils.mNM.notify(CONSTANT.PIC_TRANSMITTING_ID, mNotification);
}

}
```

(2) 注册广播。

```java
/**
 * 注册 MVR 广播器
 * 用于在线状态显示，图片传输显示
 */
private static void registerMVRBroadreceiverCast()
{
```

```
            mvrBroadcastReceiver=new MVRBroadcastReceiver();
            IntentFilter filter = new IntentFilter(CONSTANT.INLINE_STATUS_CMD);
            ctx.registerReceiver(mvrBroadcastReceiver, filter);
        }
```

(3) 注销广播。

```
    /**
     * 注销 MVR 广播器
     */
    private static void unregisterMVRBroadreceiverCast()
    {
            ctx.unregisterReceiver(mvrBroadcastReceiver);
    }
```

(4) 在线通知，传输图片通知。

```
    /**
     * 非在线提示(下线)
     */
    public static void unInlineNotification()
    {
        /*
            Intent intent=new Intent(CONSTANT.INLINE_STATUS_CMD);
            Bundle bundle=new Bundle();
            bundle.putString(CONSTANT.MVR_BROADCAST_CMD, "uninline");
            intent.putExtras(bundle);
            ctx.sendBroadcast(intent);
        */
        mNM.cancel(CONSTANT.INLINE_STATUS_ID);
    }

    /**
     * 通知栏提示图片正在传输
     */
    public static void picTransmittingNotification()
    {
        Intent intent=new Intent(CONSTANT.INLINE_STATUS_CMD);
        Bundle bundle=new Bundle();
        bundle.putString(CONSTANT.MVR_BROADCAST_CMD, "picTransmitting");
        intent.putExtras(bundle);
        ctx.sendBroadcast(intent);
    }
```

【拓展作业】

1. 参考上机部分自定义一个 BroadcastReceiver。
2. 编写实例证明 BroadcastReceiver 如何进行广播的注册与注销。

单元十
Android 中的数据存储

 课程目标

- File 文件存储
- SharedPreferences
- SQLite 的应用

简介

众所周知,数据存储是应用程序最基本的问题,任何企业系统、应用软件都必须解决这一问题,数据存储必须以某种方式保存,不能丢失,并且能够有效、简单地使用和更新这些数据。在 Android 中数据存储有五种:File 文件存储、SharedPreferences、SQLite 数据库、Contentprovider 内容提供者及网络编程。本单元将详细介绍前三种数据存储的方式。

10.1 File 文件存储

10.1.1 Android 文件的操作模式

学过 Java 的同学都知道,新建文件后就可以写入数据了,但是 Android 却不一样,因为 Android 是基于 Linux 的,我们在读写文件的时候,还需加上文件的操作模式。Android 中的操作模式如图 10-1 所示。

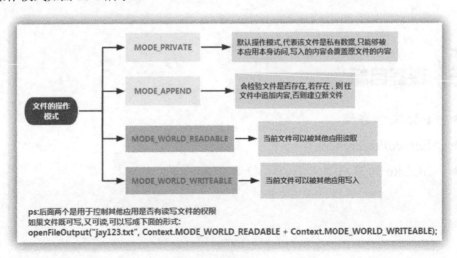

图 10-1

10.1.2 文件操作常用的 XML 属性

文件操作常用的 XML 属性及说明如表 10-1 所示。

表 10-1

XML 属性	说明
OpenFileOutput(filename,mode)	打开文件输出流,第二个参数是模式(参考图 10-1)
OpenFileInput(filename,mode)	打开文件输入流,第二个参数是模式
GetDir(name,mode)	在 APP 的 data 目录获取或创建 name 对应的子目录

(续表)

XML 属性	说明
GetFileDir()	获取 APP 的 data 目录下的 file 绝对路径
String[] fileList()	返回 app 目录下的全部文件
deleteFile(filename)	删除 app 目录下的指定文件

10.1.3 文件读写的案例实现

文件读写的案例效果如图 10-2 所示。

图 10-2

操作步骤如下。

(1) 写一个布局文件：main_activity.xml。

```xml
<LinearLayout xmlns:android="http://schemas.android.com/apk/res/android"
    xmlns:tools="http://schemas.android.com/tools"
    android:id="@+id/LinearLayout1"
    android:layout_width="match_parent"
    android:layout_height="match_parent"
    android:orientation="vertical"
    tools:context="com.jay.example.filedemo1.MainActivity">

<TextView
        android:layout_width="wrap_content"
        android:layout_height="wrap_content"
        android:text="@string/nametitle" />

<EditText
        android:id="@+id/editname"
        android:layout_width="match_parent"
        android:layout_height="wrap_content" />
```

```xml
<TextView
        android:layout_width="wrap_content"
        android:layout_height="wrap_content"
        android:text="@string/detailtitle" />

<EditText
        android:id="@+id/editdetail"
        android:layout_width="match_parent"
        android:layout_height="wrap_content"
        android:minLines="2" />

<LinearLayout
        android:layout_width="fill_parent"
        android:layout_height="wrap_content"
        android:orientation="horizontal">

<Button
        android:id="@+id/btnsave"
        android:layout_width="wrap_content"
        android:layout_height="wrap_content"
        android:text="@string/btnwrite" />

<Button
        android:id="@+id/btnclean"
        android:layout_width="wrap_content"
        android:layout_height="wrap_content"
        android:text="@string/btnclean" />
</LinearLayout>

<Button
        android:id="@+id/btnread"
        android:layout_width="wrap_content"
        android:layout_height="wrap_content"
        android:text="@string/btnread" />

</LinearLayout>
```

(2) 写一个文件协助类：FileHelper.java。

```java
public class FileHelper {

    private Context mContext;

    public FileHelper() {
    }

    public FileHelper(Context mContext) {
```

```java
        super();
        this.mContext = mContext;
    }

    /*
     * 这里定义的是一个文件保存的方法，写入文件中，所以是输出流
     * */
    public void save(String filename, String filecontent) throws Exception {
        //这里我们使用私有模式，创建出来的文件只能被本应用访问，还会覆盖原文件
        FileOutputStream output = mContext.openFileOutput(filename, Context.MODE_PRIVATE);
        output.write(filecontent.getBytes());    //将 String 字符串以字节流的形式写入输出流中
        output.close();            //关闭输出流
    }


    /*
     * 这里定义的是文件读取的方法
     * */
    public String read(String filename) throws IOException {
        //打开文件输入流
        FileInputStream input = mContext.openFileInput(filename);
        byte[] temp = new byte[1024];
        StringBuilder sb = new StringBuilder("");
        int len = 0;
        //读取文件内容:
        while ((len = input.read(temp)) > 0) {
            sb.append(new String(temp, 0, len));
        }
        //关闭输入流
        input.close();
        return sb.toString();
    }
}
```

(3) 创建一个与 MainActivity.java 关联的 main_activity.xml 布局文件，并在这里完成相关逻辑操作。

```java
public class MainActivity extends AppCompatActivity implements View.OnClickListener {

    private EditText editname;
    private EditText editdetail;
    private Button btnsave;
    private Button btnclean;
    private Button btnread;
    private Context mContext;

    @Override
    protected void onCreate(Bundle savedInstanceState) {
```

```java
        super.onCreate(savedInstanceState);
        setContentView(R.layout.activity_main);
        mContext = getApplicationContext();
        bindViews();
    }


    private void bindViews() {
        editdetail = (EditText) findViewById(R.id.editdetail);
        editname = (EditText) findViewById(R.id.editname);
        btnclean = (Button) findViewById(R.id.btnclean);
        btnsave = (Button) findViewById(R.id.btnsave);
        btnread = (Button) findViewById(R.id.btnread);

        btnclean.setOnClickListener(this);
        btnsave.setOnClickListener(this);
        btnread.setOnClickListener(this);
    }


    @Override
    public void onClick(View v) {
        switch (v.getId()) {
            case R.id.btnclean:
                editdetail.setText("");
                editname.setText("");
                break;
            case R.id.btnsave:
                FileHelper fHelper = new FileHelper(mContext);
                String filename = editname.getText().toString();
                String filedetail = editdetail.getText().toString();
                try {
                    fHelper.save(filename, filedetail);
                    Toast.makeText(getApplicationContext(), "数据写入成功", Toast.LENGTH_SHORT).show();
                } catch (Exception e) {
                    e.printStackTrace();
                    Toast.makeText(getApplicationContext(), "数据写入失败", Toast.LENGTH_SHORT).show();
                }
                break;
            case R.id.btnread:
                String detail = "";
                FileHelper fHelper2 = new FileHelper(getApplicationContext());
                try {
                    String fname = editname.getText().toString();
                    detail = fHelper2.read(fname);
```

```
                } catch (IOException e) {
                    e.printStackTrace();
                }
                Toast.makeText(getApplicationContext(), detail, Toast.LENGTH_SHORT).show();
                break;
            }
        }
    }
```

下面我们打开 DDMS 的 File Exploer 可以看到，在 data/data/<包名>/file 中有我们写入的文件，如图 10-3(a)和(b)所示。

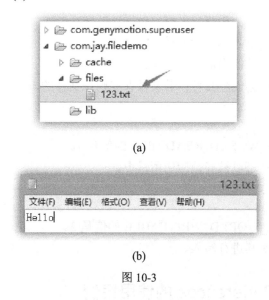

(a)

(b)

图 10-3

10.2 SharedPreferences

10.2.1 SharedPreferences 简介

　　SharedPreferences 是 Android 平台上一个轻量级的存储类，用来保存应用的一些常用配置，如 Activity 状态、Activity 暂停时等，将此 Activity 的状态保存到 SharedPreferences 中；当 Activity 重载，系统回调方法 onSaveInstanceState()时，再从 SharedPreferences 中将值取出。

　　其中的原理是，通过 Android 系统生成一个 xml 文件，并保存到/data/data/包名/shared_prefs 目录下，类似键值对的方式来存储数据。

　　SharedPreferences 提供了常规的数据类型保存接口，如 int、long、boolean、String、Float、Set 和 Map 等数据类型。

10.2.2 SharedPreferences 的使用模式

SharedPreferences 有以下 4 种使用模式。

1. 私有模式

- Context.MODE_PRIVATE 的值是 0。
- 能被创建这个文件的当前应用访问。
- 若文件不存在则会创建文件；若创建的文件已存在则会覆盖原来的文件。

2. 追加模式

- Context.MODE_APPEND 的值是 32768。
- 能被创建这个文件的当前应用访问。
- 若文件不存在则会创建文件；若文件存在则在文件的末尾进行追加内容。

3. 可读模式

- Context.MODE_WORLD_READABLE 的值是 1。
- 创建出来的文件可以被其他应用所读取。

4. 可写模式

- Context.MODE_WORLD_WRITEABLE 的值是 2。
- 允许其他应用对其进行写入。

10.2.3 SharedPreferences 的使用案例

SharedPreferences 的案例效果如图 10-4 所示。

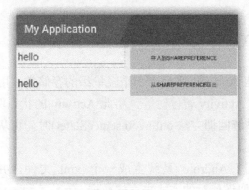

图 10-4

操作步骤如下。

(1) 新建布局文件：activity_main.xml。

`<?xml version="1.0" encoding="utf-8"?>`

```xml
<LinearLayout xmlns:android="http://schemas.android.com/apk/res/android"
    xmlns:tools="http://schemas.android.com/tools"
    android:id="@+id/activity_share_preference"
    android:layout_width="match_parent"
    android:layout_height="match_parent"
    android:orientation="vertical"
    tools:context="demo.liuchen.com.welcomepager.SharedPreferenceActivity">
<LinearLayout
        android:layout_width="match_parent"
        android:layout_height="wrap_content"
        android:orientation="horizontal">
<EditText
            android:id="@+id/edit1"
            android:layout_width="0dp"
            android:layout_weight="1"
            android:layout_height="wrap_content"
            android:hint="请输入需要的内容"/>
<Button
            android:layout_weight="1"
            android:id="@+id/btn_Save"
            android:layout_width="0dp"
            android:layout_height="wrap_content"
            android:text="存入 sharedpreference"
            android:layout_marginLeft="5dp"
            android:textSize="10dp"
            android:onClick="Click"/>

</LinearLayout>

<LinearLayout
        android:layout_width="match_parent"
        android:layout_height="wrap_content"
        android:orientation="horizontal">
<EditText
            android:id="@+id/edit2"
            android:layout_width="0dp"
            android:layout_weight="1"
            android:layout_height="wrap_content"
            android:hint="请输入需要的内容"/>
<Button
            android:layout_weight="1"
            android:id="@+id/btn_Get"
            android:layout_width="0dp"
            android:layout_height="wrap_content"
            android:text="从 sharedpreference 取出"
            android:layout_marginLeft="5dp"
            android:onClick="Click"
```

```xml
                    android:textSize="10dp"/>

</LinearLayout>
</LinearLayout>
```

(2) 完成 SharedPreferenceActivity 逻辑代码。

```java
public class SharedPreferenceActivity extends AppCompatActivity {
    private EditText meditText1 ,meditText2 ;
    private Button SaveBtn,GetBtn;
    //声明 sharedpreferenced 对象
    private SharedPreferences sp ;


    @Override
    protected void onCreate(Bundle savedInstanceState) {
        super.onCreate(savedInstanceState);
        setContentView(R.layout.activity_share_preference);

        meditText1= (EditText) findViewById(R.id.edit1);
        meditText2 = (EditText) findViewById(R.id.edit2);

        SaveBtn = (Button) findViewById(R.id.btn_Save);
        GetBtn = (Button) findViewById(R.id.btn_Get);
    }

    public void Click(View view) {
        /**
         * 获取 SharedPreferenced 对象
         * 第一个参数是生成 xml 的文件名
         * 第二个参数是存储的格式(**注意**本文后面会讲解)
         */
        sp = getSharedPreferences("User", Context.MODE_PRIVATE);
        switch (view.getId()){
            case R.id.btn_Save:
                //获取到 edit 对象
                SharedPreferences.Editor edit = sp.edit();
                //通过 editor 对象写入数据
                edit.putString("Value",meditText1.getText().toString().trim());
                //提交数据存入 xml 文件中
                edit.commit();
                break;
            case R.id.btn_Get:
                //取出数据,第一个参数是写入时的键,第二个参数是如果没有获取到数据就默认
                    返回的值
                String value = sp.getString("Value","Null");
                meditText2.setText(value);
                break;
```

```
        }
    }
}
```

10.3 SQLite 数据库

10.3.1 SQLite 简介

SQLite 是一款轻型的数据库，是遵守 ACID 的关联式数据库管理系统，它的设计目标是嵌入式的，而且目前已经在很多嵌入式产品中使用。它占用资源非常低，在嵌入式设备中，可能只需要几百 KB 的内存。它能够支持 Windows/Linux/UNIX 等主流的操作系统；同时能够与很多程序语言相结合，如 TCL、C#、PHP、Java，以及 ODBC 接口，同样比起 MySQL、PostgreSQL 这两款开源的世界著名的数据库管理系统来讲，它的处理速度更快。SQLite 第一个 Alpha 版本诞生于 2000 年 5 月，如今，SQLite 3 已经发布。

SQLite 有以下几种特性。

- 轻量级。SQLite 和 C/S 模式的数据库软件不同，它是进程内的数据库引擎，因此不存在数据库的客户端和服务器。使用 SQLite 一般只需带上它的一个动态库，就可以享受它的全部功能，而且这个动态库也相当小。
- 独立性。SQLite 数据库的核心引擎本身不依赖第三方软件，使用它也不需要安装，所以在部署的时候能够省去很多麻烦。
- 隔离性。SQLite 数据库中的所有信息都包含在一个文件内，方便管理和维护。
- 跨平台。SQLite 数据库支持大部分操作系统，如 Android、Windows Mobile、Symbin、Palm 等。
- 多语言接口。SQLite 数据库支持很多语言编程接口，如 C/C++、Java、Python、dotNet等，受到很多开发者的喜爱。
- 安全性。SQLite 数据库通过数据库级上的独占性和共享锁来实现独立事务处理。这意味着多个进程可以在同一时间从同一数据库读取数据，但只有一个可以写入数据。在某个进程或线程向数据库执行写入操作之前，必须获得独占锁定，其他读或写操作将不再发生。

10.3.2 SQLite 数据库的基础知识

1. SQLite 的数据类型

SQLite 允许忽略数据类型，但是仍然建议在 Create Table 语句中指定数据类型，因为数据类型对于我们与其他程序员之间的交流，或者准备换掉数据库引擎时能起到一个提示或帮助的作用。SQLite 支持的常见数据类型，如表 10-2 所示。

表 10-2

序 号	类型及长度
1	VARCHAR(10)
2	NVARCHAR(15)
3	TEXT
4	INTEGER
5	FLOAT
6	BOOLEAN
7	CLOB
8	BLOB
9	TIMESTAMP
10	NUMERIC(10,5)
11	VARYING CHARACTER (24)
12	NATIONAL VARYING CHARACTER(16)

2. SQLite 常用的类

SQLite 有以下几个常用的类。

- SQLiteOpenHelper：抽象类。我们通过继承该类，可以重写数据库创建及更新的方法，还可以通过该类的对象获得数据库实例或关闭数据库。
- SQLiteDatabase：数据库访问类。我们可以通过该类的对象来对数据库做一些增、删、改、查的操作。
- Cursor：游标。该类有点类似于 JDBC 中的 resultset(结果集)，可以简单理解为指向数据库中某一个记录的指针。

10.3.3 SQLite 创建、打开数据库及表

我们一起回顾一下，在 Java 应用程序中是如何处理动态数据的呢？大家都知道，Java 应用程序是通过中间件去连接另一套专业数据库软件的，如 SQL Server、Oracle 等数据库。我们也知道，Java 应用程序要连接数据库，一般在中间件的支持下写一个连接类即可。Java 连接 Oracle 10g 代码如下：

```
package com.hpsvse.util;
import java.sql.Connection;
import java.sql.DriverManager;
public class DBConnection
{
    private String driver = "oracle.jdbc.driver.OracleDriver";
    private String url = "jdbc:oracle:thin:@localhost:1521:ORCL";
```

```java
        private String user = "SCOTT";
        private String password = "TIGER";
        private Connection conn = null;
        public Connection getConnection()
        {
            try
            {
                Class.forName(driver);
                conn=DriverManager.getConnection(url, user, password);
            }
            catch (Exception e)
            {
                e.printStackTrace();
            }
            return conn;
        }
}
```

在 Android 中是如何操作 SQLite 数据库的呢？道理一样，如果我们写好的 Android 应用程序中有些数据要存放于数据库中，首先必须创建数据库和表，其次操作相应的 SQL 语句即可。在 Android 应用程序中操作嵌入式数据库 SQLite 非常简单。

通常情况下，Android 提供了一个专门操作 SQLite 数据的抽象类，我们继承它来创建数据库和表。

创建一个 DBConnection.java 类，代码如下。

```java
package com.hpsvse.util;
import android.content.Context;
import android.database.DatabaseErrorHandler;
import android.database.sqlite.SQLiteDatabase;
import android.database.sqlite.SQLiteDatabase.CursorFactory;
import android.database.sqlite.SQLiteOpenHelper;
public class DBConnection extends SQLiteOpenHelper
{
    //定义要创建的数据库
    private static String dbName = "test.db";
    //定义数据库的版本号
    private static int dbVersion = 1;
    //定义一个创建表的语句
        private static String createTable = "create table userinfo(userid integer primary key autoincrement,user_name varchar(30),user_psw varchar(20))";
    /****
     * 构造方法创建数据库
     */
    public DBConnection(Context context)
```

```
        {
            super(context, dbName, null, dbVersion);
        }
        /*********
         * 创建表
         */
        @Override
        public void onCreate(SQLiteDatabase db)
        {
            db.execSQL(createTable);
        }
        /***
         * 升级数据库
         */
        @Override
        public void onUpgrade(SQLiteDatabase db, int arg1, int arg2)
        {
            db.execSQL("DROP TABLE IF EXISTS userinfo");
            onCreate(db);
        }
    }
```

10.3.4　SQLite 案例实现

SQLite 的案例效果如图 10-5～图 10-9 所示。

图 10-5

图 10-6

图 10-7　　　　　　　　图 10-8　　　　　　　　图 10-9

代码实现步骤如下。

(1) 创建自定义类继承 SQLiteOpenHelper，实现两个方法和一个构造方法。

```java
//创建自定义类继承 SQLiteOpenHelper，实现两个方法和一个构造方法
public class MyDataBaseHelper extends SQLiteOpenHelper{
    //创建 stu 表的语句
    public static final String CREATE_BOOK = "create table stu(" +
            "id integer primary key autoincrement," +
            "name text," +
            "gender text," +
            "score real," +
            "className text)";
    //构造方法
    private Context context;
    public MyDataBaseHelper(Context context, String name, SQLiteDatabase.CursorFactory factory, int version) {
        super(context, name, factory, version);
        this.context = context;
    }
    //重写 onCreate()方法，执行建表语句
    @Override
    public void onCreate(SQLiteDatabase db) {
        //创建数据库
        db.execSQL(CREATE_BOOK);
        Toast.makeText(context,"Create successful",Toast.LENGTH_SHORT).show();

    }
    //重写 onUpgrade()方法，该方法在更新数据表时用到，此处不用
    @Override
    public void onUpgrade(SQLiteDatabase db, int oldVersion, int newVersion) {
```

            }
}

(2) 修改布局文件，添加控件，设置 ID。

```xml
<LinearLayout xmlns:android="http://schemas.android.com/apk/res/android"
    xmlns:app="http://schemas.android.com/apk/res-auto"
    xmlns:tools="http://schemas.android.com/tools"
    android:layout_width="match_parent"
    android:layout_height="match_parent"
    android:orientation="vertical"
    tools:context="com.example.myapplication2.MainActivity">

    <Button
        android:id="@+id/create_db_btn"
        android:layout_width="match_parent"
        android:layout_height="wrap_content"
        android:text="创建数据库" />

    <EditText
        android:id="@+id/main_et"
        android:layout_width="match_parent"
        android:layout_height="50dp" />

    <EditText
        android:id="@+id/main_updata_et"
        android:layout_width="match_parent"
        android:layout_height="50dp" />

    <Button
        android:id="@+id/main_insert_btn"
        android:layout_width="match_parent"
        android:layout_height="50dp"
        android:text="增" />

    <Button
        android:id="@+id/main_update_btn"
        android:layout_width="match_parent"
        android:layout_height="50dp"
        android:text="改" />

    <Button
        android:id="@+id/main_delete_btn"
        android:layout_width="match_parent"
        android:layout_height="50dp"
        android:text="删" />

    <Button
```

```
            android:id="@+id/main_show_btn"
            android:layout_width="match_parent"
            android:layout_height="wrap_content"
            android:text="查" />

</LinearLayout>
```

(3) 在 MainActivity 界面中实现对控件的监听，实现创建数据库，并进行增、删、改、查操作。

```java
public class MainActivity extends AppCompatActivity implements View.OnClickListener {

    private MyDataBaseHelper dbHelpter;
    private Button dbCreateBtn;
    private EditText editText;
    private EditText updataEt;
    private Button updataBtn;
    private Button insertBtn;
    private Button showBtn;
    private Button deleteBtn;
    private String str;
    private String TAG = "MainActivity";
    private String upstr;

    @Override
    protected void onCreate(Bundle savedInstanceState) {
        super.onCreate(savedInstanceState);
        setContentView(R.layout.activity_main);
        //创建 MyDataBaseHelper 对象(params1：上下文环境；params2：数据库名；params3：允许
        //  我们在查询数据时返回一个 cursor，这里添加 null 即可；params4：数据库版本号)
        dbHelpter = new MyDataBaseHelper(MainActivity.this, "studentdb", null, 1);
        bangID();

    }

    private void bangID() {
        dbCreateBtn = findViewById(R.id.create_db_btn);
        editText = findViewById(R.id.main_et);
        showBtn = findViewById(R.id.main_show_btn);
        insertBtn = findViewById(R.id.main_insert_btn);
        updataEt = findViewById(R.id.main_updata_et);
        updataBtn = findViewById(R.id.main_update_btn);
        deleteBtn = findViewById(R.id.main_delete_btn);

        dbCreateBtn.setOnClickListener(this);
        editText.setOnClickListener(this);
        showBtn.setOnClickListener(this);
        insertBtn.setOnClickListener(this);
```

```java
            updataBtn.setOnClickListener(this);
            deleteBtn.setOnClickListener(this);
    }

    @Override
    public void onClick(View v) {
        switch (v.getId()) {
            case R.id.main_et:
                break;
            case R.id.main_insert_btn:
                //获得输入的数据
                str = editText.getText().toString();
                //创建 SQLiteDatabase 的对象
                SQLiteDatabase db = dbHelpter.getWritableDatabase();
                //创建 ContentValues 对象，再次添加数据
                ContentValues values = new ContentValues();
                values.put("name",str);
                //插入数据(params1：表名；params2：在未指定数据的情况下赋值为空；
                // params3：值)
                db.insert("stu",null,values);
                Toast.makeText(MainActivity.this,"增成功",Toast.LENGTH_SHORT).show();
                break;
            case R.id.main_show_btn:
                //创建 SQLiteDatabase 对象
                SQLiteDatabase db1 = dbHelpter.getWritableDatabase();
                //使用游标，游历表中所有数据(params1：表名；params2：指定查询的列名；
                // params3：指定 where 约束条件；params4：占位符具体值；下面的参数与分组与
                // 排序有关)
                Cursor cursor = db1.query("stu",null,null,null,null,null,null);
                if(cursor.moveToFirst()){
                    do{
                        //取数据
                        String name = cursor.getString(cursor.getColumnIndex("name"));
                        Log.e(TAG, "onClick: "+name);
                    }while (cursor.moveToNext());
                }
                cursor.close();
                Toast.makeText(MainActivity.this,"查成功",Toast.LENGTH_SHORT).show();
                break;
            case R.id.create_db_btn:
                //创建数据库
                dbHelpter.getWritableDatabase();
                break;
            case R.id.main_update_btn:
                str = editText.getText().toString();
                upstr = updataEt.getText().toString();
                SQLiteDatabase dbupdata =dbHelpter.getWritableDatabase();
```

```
                ContentValues values1 = new ContentValues();
                values1.put("name",upstr);
                dbupdata.update("stu",values1,"name = ?",new String[]{str});
                Toast.makeText(MainActivity.this,"改成功",Toast.LENGTH_SHORT).show();
                break;
            case R.id.main_delete_btn:
                str = editText.getText().toString();
                SQLiteDatabase dbdelete =dbHelpter.getWritableDatabase();
                dbdelete.delete("stu","name=?",new String[] {str});
                Toast.makeText(MainActivity.this,"删成功",Toast.LENGTH_SHORT).show();
                break;
            default:
        }
    }
}
```

【单元小结】

- File 文件存储。
- SharedPreferences 的用法。
- SQLite 在开发中的应用。

【单元自测】

1. 在 Android 中使用 SQLiteOpenHelper 辅助类时，可以生成一个数据库，并可以对数据库版本进行管理的方法是(　　)。

    A. getWriteableDatabase()　　　　B. getReadableDatabase()

    C. getDatabase()　　　　　　　　D. getAbleDatabase()

2. SQLite 支持数据库的大小为(　　)。

    A. 2TB　　　　　　　　　　　　B. 1TB

    C. 2048M　　　　　　　　　　　D. 1000M

3. Android 提供了一个专门操作 SQLite 数据的抽象类，这个类是(　　)。

    A. SQLiteOpenHelper　　　　　　B. SQLiteOpen

    C. SQLiteHelper　　　　　　　　D. OpenHelper

4. 以下不属于 SQLite 数据类型的是(　　)。

    A. VARCHAR　　　　　　　　　B. INTEGER

    C. TEXT　　　　　　　　　　　D. VARCHAR2

5. 以下不属于 SQLite 支持的 SQL 语句的是(　　)。

    A. BEGIN TRANSACTION　　　　B. CREATE VIEW

    C. DROP INDEX　　　　　　　　D. CREATE PACKAGE

【上机实战】

上机目标

- 理解 Android 平台上数据存储机制。
- 掌握 Android+SQLite 数据库完成数据的存储。

上机练习

利用 Android+SQLite 数据库完成注册。

【问题描述】

模拟 QQ 2011 版，开发如下功能。

(1) QQ 2011 版的主界面如图 10-10 所示。

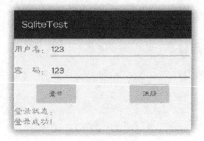

图 10-10

(2) 单击"确定"按钮后将数据存储到数据库中。

【问题分析】

使用 Android 完成表示层的工作。

【参考步骤】

(1) 创建 Android 工程。
(2) 开发主页面 MainActivity.java，具体代码如下。
OpenHelper 类：

```
public class OpenHelper extends SQLiteOpenHelper {

    //建表语句
    public static final String CREATE_USER = "create table User ("
            + "id integer primary key autoincrement, "
            + "username text, "
            + "userpwd text)";

    public OpenHelper(Context context, String name, CursorFactory factory,
```

```java
                    int version) {
        super(context, name, factory, version);
        // TODO Auto-generated constructor stub
    }

    @Override
    public void onCreate(SQLiteDatabase db) {
        // TODO Auto-generated method stub

        db.execSQL(CREATE_USER);//创建用户表
    }

    @Override
    public void onUpgrade(SQLiteDatabase db, int oldVersion, int newVersion) {
        // TODO Auto-generated method stub

    }

}
```

SqliteDB 类:

```java
public class SqliteDB {
    /**
     * 数据库名
     */
    public static final String DB_NAME = "sqlite_dbname";
    /**
     * 数据库版本
     */
    public static final int VERSION = 1;

    private static SqliteDB sqliteDB;

    private SQLiteDatabase db;

    private SqliteDB(Context context) {
        OpenHelper dbHelper = new OpenHelper(context, DB_NAME, null,VERSION);
        db = dbHelper.getWritableDatabase();
    }

    /**
     * 获取 SqliteDB 实例
     * @param context
     */
    public synchronized static SqliteDB getInstance(Context context) {
        if (sqliteDB == null) {
            sqliteDB = new SqliteDB(context);
```

```java
            }
            return sqliteDB;
    }

    /**
     * 将 User 实例存储到数据库。
     */
    public int  saveUser(User user) {
        if (user != null) {
            /* ContentValues values = new ContentValues();
               values.put("username", user.getUsername());
               values.put("userpwd", user.getUserpwd());
               db.insert("User", null, values);*/

            Cursor cursor = db.rawQuery("select * from User where username=?", new String[]{user.getUsername().toString()});
            if (cursor.getCount() > 0) {
                return -1;
            } else {
                try {
                    db.execSQL("insert into User(username,userpwd) values(?,?) ", new String[]{user.getUsername().toString(), user.getUserpwd().toString()});
                } catch (Exception e) {
                    Log.d("错误", e.getMessage().toString());
                }
                return 1;
            }
        }
        else {
            return 0;
        }
    }

    /**
     * 从数据库读取 User 信息。
     */
    public List<User> loadUser() {
        List<User> list = new ArrayList<User>();
        Cursor cursor = db
                .query("User", null, null, null, null, null, null);
        if (cursor.moveToFirst()) {
            do {
                User user = new User();
                user.setId(cursor.getInt(cursor.getColumnIndex("id")));
                user.setUsername(cursor.getString(cursor
                        .getColumnIndex("username")));
                user.setUserpwd(cursor.getString(cursor
```

```java
                            .getColumnIndex("userpwd")));
                    list.add(user);
            } while (cursor.moveToNext());
        }
        return list;
    }

    public int Quer(String pwd,String name)
    {


        HashMap<String,String> hashmap=new HashMap<String,String>();
        Cursor cursor =db.rawQuery("select * from User where username=?", new String[]{name});

       // hashmap.put("name",db.rawQuery("select * from User where name=?",new String[]{name}).toString());
        if (cursor.getCount()>0)
        {
            Cursor pwdcursor =db.rawQuery("select * from User where userpwd=? and username=?",new String[]{pwd,name});
            if (pwdcursor.getCount()>0)
            {
                return 1;
            }
            else {
                return -1;
            }
        }
        else {
            return 0;
        }



    }
}
```

User 类：

```java
package model;

/**
 * Created by Administrator on 2015/11/10.
 */
public class User {

    private int id;
    public int getId() {
```

```java
        return id;
    }

    public void setId(int id) {
        this.id = id;
    }



    public String getUsername() {
        return username;
    }

    public void setUsername(String username) {
        this.username = username;
    }

    public String getUserpwd() {
        return userpwd;
    }

    public void setUserpwd(String userpwd) {
        this.userpwd = userpwd;
    }

    private String username;
    private String userpwd;

}
```

LoginActivity：

```java
package activity;
import android.os.Bundle;
import android.support.v7.app.AppCompatActivity;
import android.view.View;
import android.widget.Button;
import android.widget.EditText;
import android.widget.TextView;

import com.example.administrator.sqlitetest.R;

import java.util.ArrayList;
import java.util.HashMap;
import java.util.List;

import model.SqliteDB;
import model.User;
```

```java
public class LoginActivity extends AppCompatActivity {
    private Button reg;
    private Button login;
    private EditText count;
    private EditText pwd;
    private TextView state;
    private List<User> userList;
    private List<User> dataList = new ArrayList<>();
    @Override
    protected void onCreate(Bundle savedInstanceState) {
        super.onCreate(savedInstanceState);
        setContentView(R.layout.activity_login);


        reg= (Button) findViewById(R.id.regin);
        login= (Button) findViewById(R.id.login);
        count= (EditText) findViewById(R.id.count);
        pwd= (EditText) findViewById(R.id.pwd);
        state= (TextView) findViewById(R.id.state);
        reg.setOnClickListener(new View.OnClickListener() {
            @Override
            public void onClick(View v) {
                String name=count.getText().toString().trim();
                String pass=pwd.getText().toString().trim();

                User user=new User();
                user.setUsername(name);
                user.setUserpwd(pass);

              int result=SqliteDB.getInstance(getApplicationContext()).saveUser(user);
                if (result==1){
                    state.setText("注册成功！");
                }else   if (result==-1)
                {
                    state.setText("用户名已经存在！");
                }
                else
                {
                    state.setText("！");
                }


            }
        });
        login.setOnClickListener(new View.OnClickListener() {
            @Override
            public void onClick(View v) {
                String name=count.getText().toString().trim();
```

```
                    String pass=pwd.getText().toString().trim();
                    //userList=SqliteDB.getInstance(getApplicationContext()).loadUser();
                    int result=SqliteDB.getInstance(getApplicationContext()).Quer(pass,name);
                    if (result==1)
                    {
                            state.setText("登录成功！");
                    }
                    else if (result==0){
                        state.setText("用户名不存在！");

                    }
                    else if(result==-1)
                    {
                        state.setText("密码错误！");
                    }
/*                    for (User user : userList) {

                            if (user.getUsername().equals(name))
                            {
                                if (user.getUserpwd().equals(pass))
                                {
                                    state.setText("登录成功！");

                                }else {
                                    state.setText("密码错误！");

                                }
                            }
                            else {
                                state.setText("用户名不存在！");

                            }
                    }*/
                }
            });
        }
    }
```

【拓展作业】

1. SQLite 数据库连接方式有哪些？
2. 写一个 Android 小程序，从 SQLite 数据库中获取数据。

单元十一 ContentProvider

 课程目标

- ► ContentProvider 概述
- ► Uri 和 UriMatcher
- ► 操作 ContentProvider
- ► 自定义 ContentProvider

简介

ContentProvider 即内容提供商或内容提供器。Android 中的 ContentProvider 机制可支持在多个应用中存储和读取数据,这也是跨应用共享数据的方式之一,还有文件、SharedPreference、SQLite 数据库等方式共享存储数据库,但是 ContentProvider 更好地提供了数据共享接口的统一性。在 Android 系统中,没有一个公共的内存区域供多个应用共享存储数据。本单元将会重点介绍数据共享机制。

11.1 ContentProvider 概述

ContentProvider在Android中的作用是对外共享数据。也就是说,我们可以通过ContentProvider把应用中的数据共享给其他应用访问,其他应用可以通过ContentProvider对应用中的数据进行添、删、改、查。关于数据共享,我们学习过文件操作模式,知道通过指定文件操作模式Context.MODE_WORLD_READABLE或Context.MODE_WORLD_WRITEABLE同样也可以对外共享数据。那么,这里为何要使用ContentProvider对外共享数据呢?因为,如果采用文件操作模式对外共享数据,那么数据的访问方式会因数据存储的方式而不同,导致数据的访问方式无法统一。例如,采用XML文件对外共享数据,需要进行XML解析才能读取数据;采用SharedPreferences共享数据,需要使用SharedPreferences API读取数据,如图11-1所示。

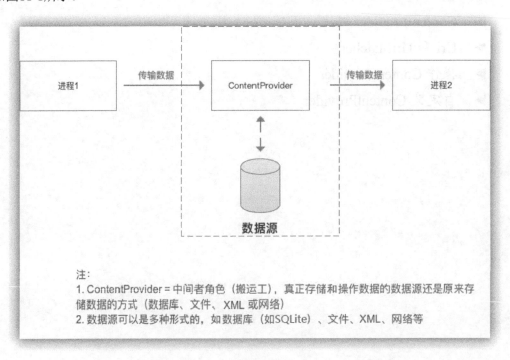

图 11-1

Android 应用继承 ContentProvider 类，并重写该类用于提供数据和存储数据的方法，即可向其他应用共享其数据。使用 ContentProvider 共享数据的好处是统一了数据访问方式。

Android 应用要通过 ContentProvider 对外共享数据时，需执行以下两个步骤。

(1) 继承 ContentProvider 并重写下面的方法(注意，以下只是部分代码)。

```
public class PersonContentProvider extends ContentProvider{
    public boolean onCreate(){}
    public Uri insert(Uri uri, ContentValues values){}
    public int delete(Uri uri, String selection, String[] selectionArgs){}
    public int update(Uri uri, ContentValues values, String selection, String[] selectionArgs){}
    public Cursor query(Uri uri, String[] projection, String selection, String[] selectionArgs, String sortOrder){}
    public String getType(Uri uri){}
}
```

(2) 在 AndroidManifest.xml 中使用<provider>对该 ContentProvider 进行配置，为了能让其他应用找到该 ContentProvider，ContentProvider 采用了 authorities(主机名/域名)对它进行唯一标识，可以把 ContentProvider 看作一个网站(网站也是数据提供者)，authorities 就是它的域名。

```
<manifest .... >
<application android:icon="@drawable/icon" android:label="@string/app_name">
<provider android:name=".PersonContentProvider"
android:authorities="com.jbridge.provider.personprovider"/>
</application>
</manifest>
```

注意

一旦应用继承了 ContentProvider 类，后面我们就会把这个应用称为 ContentProvider (内容提供者)。

11.2 Uri 和 UriMatcher

11.2.1 Uri 介绍

Uri 代表了要操作的数据，主要包含了两部分信息：①需要操作的 ContentProvider；②对 ContentProvider 中的什么数据进行操作。一个 Uri 由如图 11-2 所示的几部分组成。

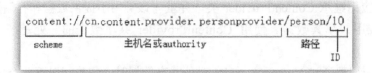

图 11-2

ContentProvider(内容提供者)的 scheme 已经由 Android 所规定，为 content://。

主机名(或叫 authority)用于唯一标识这个 ContentProvider，外部调用者可以根据该标识来找到它。路径(path)可以用来表示我们要操作的数据，路径的构建应根据业务确定，具体如下。

- 要操作 person 表中 ID 为 10 的记录，可以构建这样的路径：/person/10。
- 要操作 person 表中 ID 为 10 的记录的 name 字段，可以构建这样的路径：person/10/name。
- 要操作 person 表中的所有记录，可以构建这样的路径：/person。
- 要操作×××表中的记录，可以构建这样的路径：/×××。
- 要操作的数据不一定来自数据库，也可以是文件、XML 或网络等其他存储方式，如要操作 XML 文件中 person 节点下的 name 节点，可以构建这样的路径：/person/name。

如果要把一个字符串转换成 Uri，可以使用 Uri 类中的 parse()方法，代码如下。

Uriuri= Uri.parse("content://cn.content.provider.personprovider/person");

11.2.2 使用 UriMatcher

因为 Uri 代表了要操作的数据，所以经常需要解析 Uri，并从 Uri 中获取数据。Android 系统提供了两个用于操作 Uri 的工具类，分别为 UriMatcher 和 ContentUris，掌握它们的使用方法，会便于开发工作。UriMatcher 类用于匹配 Uri，它的用法如下。

(1) 注册需要匹配的 Uri 路径，代码如下。

```
//常量 UriMatcher.NO_MATCH 表示不匹配任何路径的返回码
UriMatcher    sMatcher = new UriMatcher(UriMatcher.NO_MATCH);
//如果 match()方法匹配 content://cn.content.provider.personprovider/person 路径，返回匹配码为 1
sMatcher.addURI( " cn.content.provider.personprovider " , " person " ,1);//添加需要匹配的 Uri，如果匹配
//就会返回匹配码
//如果 match()方法匹配 content://cn.content.provider.personprovider/person/230 路径，返回匹配码为 2
sMatcher.addURI( " cn.content.provider.personprovider " , " person/# " , 2);//#号为通配符
switch (sMatcher.match(Uri.parse("content://cn.content.provider.personprovider/person/10"))) {
    case 1
      break;
    case 2
      break;
  default://不匹配
```

```
        break;
}
```

(2) 注册需要匹配的 Uri 后，可以使用 sMatcher.match(uri)方法对输入的 Uri 进行匹配，如果匹配就返回匹配码，匹配码是调用 addURI()方法传入的第三个参数，若匹配 content://cn.content.provider.personprovider/person 路径，则返回的匹配码为 1。

11.2.3 ContentUris 类及数据的共享

ContentUris 类用于获取 Uri 路径后面的 ID 部分，它有以下两个比较实用的方法。
(1) withAppendedId(uri,id)方法：用于为路径加上 ID 部分，代码如下。

```
Uri uri = Uri.parse("content://cn.content.provider.personprovider/person")
Uri resultUri = ContentUris.withAppendedId(uri, 10);
//生成后的 Uri 为：content://cn.content.provider.personprovider/person/10
```

(2) parseId(uri)方法：用于从路径中获取 ID 部分，代码如下。

```
Uri uri = Uri.parse("content://cn.content.provider.personprovider/person/10")
long personid = ContentUris.parseId(uri);//获取的结果为 10
```

ContentProvider 类主要方法的作用如下。

- publicbooleanonCreate()。该方法在 ContentProvider 创建后就会被调用，Android 开机后，ContentProvider 在其他应用第一次访问它时才会被创建。
- publicUriinsert(Uri uri, ContentValuesvalues)。该方法用于供外部应用往 ContentProvider 中添加数据。
- publicintdelete(Uri uri,String selection, String[] selectionArgs)。该方法用于供外部应用从 ContentProvider 中删除数据。
- publicintupdate(Uri uri,ContentValuesvalues, String selection, String[] selectionArgs)。该方法用于供外部应用更新 ContentProvider 中的数据。
- publicCursorquery(Uri uri,String[] projection, String selection, String[] selectionArgs, StringsortOrder)。该方法用于供外部应用从 ContentProvider 中获取数据。
- publicStringgetType(Uriuri)。该方法用于返回当前 Uri 所代表数据的 MIME 类型。如果操作的数据属于集合类型，那么 MIME 类型字符串应该以 vnd.android. cursor.dir/开头，例如：要得到所有 person 记录的 Uri 为 content://cn.content.provider. personprovider/person，那么返回的 MIME 类型字符串应为 vnd.android.cursor. dir/person。如果要操作的数据属于非集合类型数据，那么 MIME 类型字符串应以 vnd.android.cursor.item/开头，例如：得到 ID 为 10 的 person 记录,Uri 为 content://cn. content.provider.personprovider/person/10，那么返回的 MIME 类型字符串应为 vnd.android.cursor.item/person。

11.3 操作 ContentProvider

11.3.1 使用 ContentResolver 操作 ContentProvider 中的数据

当外部应用需要对 ContentProvider 中的数据进行添加、删除、修改和查询操作时，可以使用 ContentResolver 类来完成，要获取 ContentResolver 对象，可以使用 Activity 提供的 getContentResolver()方法。ContentResolver 类提供了与 ContentProvider 类相同签名的 4 个方法，具体如下：

- publicUriinsert(Uri uri, ContentValuesvalues)。该方法用于往 ContentProvider 中添加数据。
- publicintdelete(Uri uri, String selection, String[] selectionArgs)。该方法用于从 ContentProvider 中删除数据。
- publicintupdate(Uri uri, ContentValuesvalues, String selection, String[] selectionArgs)。该方法用于更新 ContentProvider 中的数据。
- publicCursorquery(Uri uri, String[] projection, String selection, String[] selectionArgs, String sortOrder)。该方法用于从 ContentProvider 中获取数据。

这些方法的第一个参数为 Uri，代表要操作的 ContentProvider 和对其中的什么数据进行操作，假设给定的是 Uri.parse("content://cn.content.providers.personprovider/person/10")，那么将会对主机名为 cn.content.providers.personprovider 的 ContentProvider 进行操作，操作的数据为 person 表中 ID 为 10 的记录。

使用 ContentResolver 对 ContentProvider 中的数据进行添加、删除、修改和查询操作，代码如下。

```
ContentResolver resolver =   getContentResolver();
Uri uri = Uri.parse("content://cn.content.provider.personprovider/person");
//添加一条记录
ContentValues values = new ContentValues();
values.put("name", "itcast");
values.put("age", 25);
resolver.insert(uri, values);
//获取 person 表中所有记录
Cursor cursor = resolver.query(uri, null, null, null, "personid desc");
while(cursor.moveToNext()){
    Log.i("ContentTest", "personid="+ cursor.getInt(0)+ ",name="+ cursor.getString(1));
}
//把 ID 为 1 的记录的 name 字段值更新为 liming
ContentValues updateValues = new ContentValues();
updateValues.put("name", "liming");
Uri updateIdUri = ContentUris.withAppendedId(uri, 2);
resolver.update(updateIdUri, updateValues, null, null)
```

```
//删除 ID 为 2 的记录
Uri deleteIdUri = ContentUris.withAppendedId(uri, 2);
resolver.delete(deleteIdUri, null, null);
```

11.3.2 通讯录操作

使用 ContentResolver 对通讯录中的数据进行添加、删除、修改和查询操作。
加入读写联系人信息的权限，代码如下。

```
<uses-permission android:name="android.permission.READ_CONTACTS"/>
<uses-permission android:name="android.permission.WRITE_CONTACTS"/>
```

加入读取联系人信息的权限，代码如下。

```
<uses-permission android:name="android.permission.READ_CONTACTS"/>
```
content://com.android.contacts/contacts 操作的数据是联系人信息 Uri
content://com.android.contacts/data/phones 联系人电话 Uri
content://com.android.contacts/data/emails 联系人 Email Uri

读取联系人信息，代码如下。

```
Cursor cursor = getContentResolver().query(ContactsContract.Contacts.CONTENT_URI,
null, null, null, null);
    while (cursor.moveToNext()) {
        String contactId = cursor.getString(cursor.getColumnIndex(ContactsContract.Contacts._ID));
        String name = cursor.getString(cursor.getColumnIndex(ContactsContract.Contacts.DISPLAY_NAME));

        Cursor phones =
getContentResolver().query(ContactsContract.CommonDataKinds.Phone.CONTENT_URI,
            null,
            ContactsContract.CommonDataKinds.Phone.CONTACT_ID +" = "+ contactId,
            null, null);
        while (phones.moveToNext()) {
            String phoneNumber = phones.getString(phones.getColumnIndex(
                ContactsContract.CommonDataKinds.Phone.NUMBER));
            Log.i("RongActivity", "phoneNumber="+phoneNumber);
        }
        phones.close();

        Cursor emails =
getContentResolver().query(ContactsContract.CommonDataKinds.Email.CONTENT_URI,
            null,
            ContactsContract.CommonDataKinds.Email.CONTACT_ID + " = " + contactId,
            null, null);
        while (emails.moveToNext()) {
            // This would allow you get several email addresses
            String emailAddress =
```

```
            emails.getString(emails.getColumnIndex(ContactsContract.CommonDataKinds.Email.DATA));
            Log.i("RongActivity", "emailAddress="+ emailAddress);
        }
        emails.close();
    }
    cursor.close();
```

添加联系人，有以下两种方法。

方法一：

```
/**
 * 首先向 RawContacts.CONTENT_URI 执行一个空值插入，目的是获取系统返回的
   rawContactId
 * 这时后面插入 data 表的依据,只有执行空值插入,才能使插入的联系人在通讯录里可见
 */
public void testInsert() {
    ContentValues values = new ContentValues();
    //首先向 RawContacts.CONTENT_URI 执行一个空值插入，目的是获取系统返回的
        rawContactId
    Uri rawContactUri = this.getContext().getContentResolver().insert(RawContacts.CONTENT_URI,
        values);
    long rawContactId = ContentUris.parseId(rawContactUri);
    //往 data 表插入姓名数据
    values.clear();
    values.put(Data.RAW_CONTACT_ID, rawContactId);
    values.put(Data.MIMETYPE, StructuredName.CONTENT_ITEM_TYPE);//内容类型
    values.put(StructuredName.GIVEN_NAME, "John");
    this.getContext().getContentResolver().insert(android.provider.ContactsContract.Data.
        CONTENT_URI,values);
    //往 data 表插入电话数据
    values.clear();
    values.put(Data.RAW_CONTACT_ID, rawContactId);
    values.put(Data.MIMETYPE, Phone.CONTENT_ITEM_TYPE);
    values.put(Phone.NUMBER, "13100000120");
    values.put(Phone.TYPE, Phone.TYPE_MOBILE);
    this.getContext().getContentResolver().insert(android.provider.ContactsContract.Data.
        CONTENT_URI,values);
    //往 data 表插入 E-mail 数据
    values.clear();
    values.put(Data.RAW_CONTACT_ID, rawContactId);
    values.put(Data.MIMETYPE, Email.CONTENT_ITEM_TYPE);
    values.put(Email.DATA, "liming@content.cn");
    values.put(Email.TYPE, Email.TYPE_WORK);
    this.getContext().getContentResolver().insert(android.provider.ContactsContract.Data.
        CONTENT_URI,values);
}
```

方法二：批量添加，处于同一个事务中。

```java
public void testSave() throws Throwable{
    //文档位置：reference\android\provider\ContactsContract.RawContacts.html
    ArrayList<ContentProviderOperation> ops = new ArrayList<ContentProviderOperation>();
    int rawContactInsertIndex = ops.size();
    ops.add(ContentProviderOperation.newInsert(RawContacts.CONTENT_URI)
            .withValue(RawContacts.ACCOUNT_TYPE, null)
            .withValue(RawContacts.ACCOUNT_NAME, null)
            .build());
    //文档位置：reference\android\provider\ContactsContract.Data.html

ops.add(ContentProviderOperation.newInsert(android.provider.ContactsContract.Data.CONTENT_URI)
            .withValueBackReference(Data.RAW_CONTACT_ID, rawContactInsertIndex)
            .withValue(Data.MIMETYPE, StructuredName.CONTENT_ITEM_TYPE)
            .withValue(StructuredName.GIVEN_NAME, "Kate")
            .build());

ops.add(ContentProviderOperation.newInsert(android.provider.ContactsContract.Data.CONTENT_URI)
            .withValueBackReference(Data.RAW_CONTACT_ID, rawContactInsertIndex)
            .withValue(Data.MIMETYPE, Phone.CONTENT_ITEM_TYPE)
            .withValue(Phone.NUMBER, "13600000001")
            .withValue(Phone.TYPE, Phone.TYPE_MOBILE)
            .withValue(Phone.LABEL, "手机号")
            .build());
    ops.add(ContentProviderOperation.newInsert(android.provider.ContactsContract.Data.
        CONTENT_URI)
            .withValueBackReference(Data.RAW_CONTACT_ID, rawContactInsertIndex)
            .withValue(Data.MIMETYPE, Email.CONTENT_ITEM_TYPE)
            .withValue(Email.DATA, "liming@content.cn")
            .withValue(Email.TYPE, Email.TYPE_WORK)
            .build());
    ContentProviderResult[] results = this.getContext().getContentResolver()
        .applyBatch(ContactsContract.AUTHORITY, ops);
    for(ContentProviderResult result : results){
        Log.i(TAG, result.uri.toString());
    }
}
```

11.4 自定义 ContentProvider

我们先来看一下，Uri 与行为代码的映射关系，如下。

MatcherConst：

```java
public class MatcherConst {
public final static String AUTHORITY="com.breakloop.contentproviderdemo1";

    public final static int BY_NAME=1;

    public final static int BY_AGE=2;

    public final static int BY_SEX=3;

    public final static int BY_NONE=0;

    public final static String PATH_BY_NAME=DBConst.TABLE_PERSON+"/ByName/*";

    public final static String PATH_BY_AGE=DBConst.TABLE_PERSON+"/ByAge/#";

    public final static String PATH_BY_SEX=DBConst.TABLE_PERSON+"/BySex/*";
}
```

UriMatcher：

```java
static UriMatcher matcher;

static {
    matcher=new UriMatcher(UriMatcher.NO_MATCH);
    matcher.addURI(MatcherConst.AUTHORITY,MatcherConst.PATH_BY_NAME,
        MatcherConst.BY_NAME);
    matcher.addURI(MatcherConst.AUTHORITY,MatcherConst.PATH_BY_AGE,
        MatcherConst.BY_AGE);
    matcher.addURI(MatcherConst.AUTHORITY,MatcherConst.PATH_BY_SEX,
        MatcherConst.BY_SEX);
    matcher.addURI(MatcherConst.AUTHORITY,DBConst.TABLE_PERSON,
        MatcherConst.BY_NONE);
}
```

在上面的示例中，UriMatch 绑定了 4 个 Uri，并将各个 Uri 映射为 4 个行为代码。

那么如何将行为代码映射为具体的数据库操作呢？换句话说，在哪儿使用 UriMatcher 呢？当然是在 ContentProvider 中！在 ContentProvider 中的增、删、改、查方法中，完成操作映射，步骤如下。

（1）用 Android Studio 创建一个 ContentProvider，如图 11-3 所示。

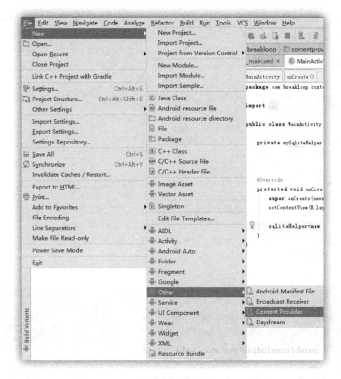

图 11-3

(2) 创建过程中，需要提供 AUTHORITY，如图 11-4 所示。

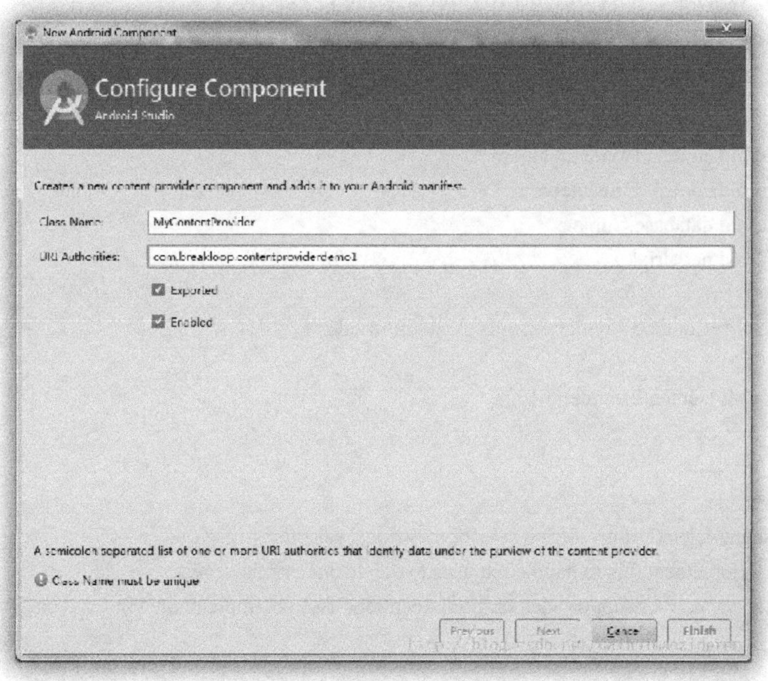

图 11-4

(3) 生成 ContentProvider 后，Android Studio 将自动帮助 ContentProvider 在 Manifest 中进行注册，如图 11-5 所示。

```xml
<application
    android:allowBackup="true"
    android:icon="@mipmap/ic_launcher"
    android:label="Contentproviderdemo1"
    android:roundIcon="@mipmap/ic_launcher_round"
    android:supportsRtl="true"
    android:theme="@style/AppTheme">
    <activity android:name=".MainActivity">
        <intent-filter>
            <action android:name="android.intent.action.MAIN" />

            <category android:name="android.intent.category.LAUNCHER" />
        </intent-filter>
    </activity>

    <provider
        android:name=".MyContentProvider"
        android:authorities="com.breakloop.contentproviderdemo1"
        android:enabled="true"
        android:exported="true"></provider>
</application>
```

图 11-5

接下来，我们看一看 ContentProvider 的结构，代码如下。

```java
package com.breakloop.contentproviderdemo1;

import android.content.ContentProvider;
import android.content.ContentValues;
import android.content.UriMatcher;
import android.database.Cursor;
import android.net.Uri;

public class MyContentProvider extends ContentProvider {

    public MyContentProvider() {
    }

    @Override
    public int delete(Uri uri, String selection, String[] selectionArgs) {
        // Implement this to handle requests to delete one or more rows.
        throw new UnsupportedOperationException("Not yet implemented");
    }

    @Override
    public String getType(Uri uri) {
```

```java
            // TODO: Implement this to handle requests for the MIME type of the data
            // at the given URI.
            throw new UnsupportedOperationException("Not yet implemented");
        }

        @Override
        public Uri insert(Uri uri, ContentValues values) {
            // TODO: Implement this to handle requests to insert a new row.
            throw new UnsupportedOperationException("Not yet implemented");
        }

        @Override
        public boolean onCreate() {
            // TODO: Implement this to initialize your content provider on startup.
            return false;
        }

        @Override
        public Cursor query(Uri uri, String[] projection, String selection,
                            String[] selectionArgs, String sortOrder) {
            // TODO: Implement this to handle query requests from clients.
            throw new UnsupportedOperationException("Not yet implemented");
        }

        @Override
        public int update(Uri uri, ContentValues values, String selection,
                          String[] selectionArgs) {
            // TODO: Implement this to handle requests to update one or more rows.
            throw new UnsupportedOperationException("Not yet implemented");
        }
    }
```

关于构造方法，此处不再赘述。

关于初始化方法，当返回 true 时，表明初始化成功，否则，失败。由于我们要对数据库进行操作，因此，需要获取 SQLite 数据库对象。

```java
private myHelper dbHelper;

public boolean onCreate() {
    helper=new mySqliteHelper(getContext(),DBConst.DB_NAME,null,1);
    return true;
}
```

关于数据库方法，我们看到传参中存在 Uri，因此，这里需要用到 UirMatcher。将刚才的 UriMatcher 代码段加入 MyContentProvider.，即可在各个数据库方法中解析 Uri。同时，由于 SQLite 数据库对象的存在，进而可以对数据库进行相应的操作。

我们先看一下最简单的插入操作，代码如下。

```java
@Override
public Uri insert(Uri uri, ContentValues values) {
    Uri returnUri=null;
    SQLiteDatabase db=helper.getWritableDatabase();
    switch (matcher.match(uri)){
        case MatcherConst.BY_NONE:
        long recordID=db.insert(DBConst.TABLE_PERSON,null,values);
        returnUri=Uri.parse("content://"+MatcherConst.AUTHORITY+"/"+DBConst.
            TABLE_PERSON+"/"+recordID);
            break;
        default:
            break;
    }

    return returnUri;
}
```

再来看一下稍微复杂的查询，代码如下。

```java
@Override
public Cursor query(Uri uri, String[] projection, String selection,
String[] selectionArgs, String sortOrder) {
    Cursor cursor=null;
    SQLiteDatabase db=helper.getReadableDatabase();
    switch (matcher.match(uri)){
        case MatcherConst.BY_NONE:
        cursor=db.query(DBConst.TABLE_PERSON,projection,selection,selectionArgs,null,null,
            sortOrder);
            break;
        case MatcherConst.BY_AGE:
        cursor=db.query(DBConst.TABLE_PERSON,projection,DBConst.COLUMN_AGE+"=?",
            new String[]{uri.getPathSegments().get(2)},null,null,sortOrder);
            break;
        case MatcherConst.BY_SEX:
        cursor=db.query(DBConst.TABLE_PERSON,projection,DBConst.COLUMN_SEX+"=?",
            new String[]{uri.getPathSegments().get(2)},null,null,sortOrder);
            break;
        case MatcherConst.BY_NAME:
        cursor=db.query(DBConst.TABLE_PERSON,projection,DBConst.COLUMN_NAME+"=?",
            new String[]{uri.getPathSegments().get(2)},null,null,sortOrder);
            break;
        default:
            break;
    }

    return cursor;
}
```

这里需要注意的是，如何取 Uri 中的传入数据。使用的获取方法是 uri.getPathSegments().get(index)。该方法获取的是 AUTHORITY 后面的 String 部分，然后，以 "/" 为分隔符，生成 String[]。

接着是更新操作，代码如下。

```java
@Override
public int update(Uri uri, ContentValues values, String selection,
String[] selectionArgs) {
    int recordID=0;
    SQLiteDatabase db=helper.getWritableDatabase();
    switch (matcher.match(uri)){
        case MatcherConst.BY_NONE:
            recordID=db.update(DBConst.TABLE_PERSON,values,null,null);
            break;
        case MatcherConst.BY_AGE:
        recordID=db.update(DBConst.TABLE_PERSON,values,DBConst.COLUMN_AGE+"=?",
            new String[]{uri.getPathSegments().get(2)});
            break;
        case MatcherConst.BY_SEX:
        recordID=db.update(DBConst.TABLE_PERSON,values,DBConst.COLUMN_SEX+"=?",
            new String[]{uri.getPathSegments().get(2)});
            break;
        case MatcherConst.BY_NAME:
        recordID=db.update(DBConst.TABLE_PERSON,values,DBConst.COLUMN_NAME+"=?",
            new String[]{uri.getPathSegments().get(2)});
            break;
        default:
            break;
    }
    return recordID;
}
```

删除操作，代码如下。

```java
@Override
public int delete(Uri uri, String selection, String[] selectionArgs) {
    int recordID=0;
    SQLiteDatabase db=helper.getWritableDatabase();
    switch (matcher.match(uri)){
        case MatcherConst.BY_NONE:
            recordID=db.delete(DBConst.TABLE_PERSON,null,null);
            break;
        case MatcherConst.BY_AGE:
        recordID=db.delete(DBConst.TABLE_PERSON,DBConst.COLUMN_AGE+"=?",
            new String[]{uri.getPathSegments().get(2)});
            break;
        case MatcherConst.BY_SEX:
        recordID=db.delete(DBConst.TABLE_PERSON,DBConst.COLUMN_SEX+"=?",
```

```
                    new String[]{uri.getPathSegments().get(2)});
                break;
            case MatcherConst.BY_NAME:
                recordID=db.delete(DBConst.TABLE_PERSON,DBConst.COLUMN_NAME+"=?",
                    new String[]{uri.getPathSegments().get(2)});
                break;
            default:
                break;
        }
    return recordID;
}
```

那么，第三方如何调用 ContentProvider 呢？

这里，我们新建一个工程 contentproviderdemo2，在必要的位置使用方法，代码如下。

```
public ContentResolver getContentResolver()
```

获取 ContentProvider 实例后，便可传入 Uri，调用数据库的相关方法。

例如，我们插入 4 条记录，代码如下。

```
Uri returnUri=null;
ContentResolver resolver=getContentResolver();
Uri uri=Uri.parse("content://"+MatcherConst.AUTHORITY+"/"+DBConst.TABLE_PERSON);
ContentValues values=new ContentValues();

values.put(DBConst.COLUMN_NAME,"A");
values.put(DBConst.COLUMN_AGE,10);
values.put(DBConst.COLUMN_SEX,"Male");
returnUri=resolver.insert(uri,values);
if(returnUri!=null)
    Log.i(TAG, "return Uri = "+returnUri.toString());

values.put(DBConst.COLUMN_NAME,"B");
values.put(DBConst.COLUMN_AGE,11);
values.put(DBConst.COLUMN_SEX,"Male");
returnUri=resolver.insert(uri,values);
if(returnUri!=null)
    Log.i(TAG, "return Uri = "+returnUri.toString());

values.put(DBConst.COLUMN_NAME,"C");
values.put(DBConst.COLUMN_AGE,12);
values.put(DBConst.COLUMN_SEX,"Female");
returnUri=resolver.insert(uri,values);
if(returnUri!=null)
    Log.i(TAG, "return Uri = "+returnUri.toString());

values.put(DBConst.COLUMN_NAME,"D");
values.put(DBConst.COLUMN_AGE,13);
values.put(DBConst.COLUMN_SEX,"Female");
```

```
returnUri=resolver.insert(uri,values);
if(returnUri!=null)
    Log.i(TAG, "return Uri = "+returnUri.toString());
```

我们看一下输出结果,如下。

```
I/com.breakloop.contentproviderdemo2.MainActivity: return Uri =
    content://com.breakloop.contentproviderdemo1/PERSON/1
I/com.breakloop.contentproviderdemo2.MainActivity: return Uri =
    content://com.breakloop.contentproviderdemo1/PERSON/2
I/com.breakloop.contentproviderdemo2.MainActivity: return Uri =
    content://com.breakloop.contentproviderdemo1/PERSON/3
I/com.breakloop.contentproviderdemo2.MainActivity: return Uri =
    content://com.breakloop.contentproviderdemo1/PERSON/4
1
```

写入成功,进行查询,如下。

```
public void selectRecord(){
    Uri uri;
    String name="A";
    int age=11;
    String sex="Male";

    Log.i(TAG, "Select by Name");
    uri=Uri.parse("content://"+MatcherConst.AUTHORITY+"/"+DBConst.TABLE_PERSON+
        "/ByName/"+name);
    selectRecord(uri);

    Log.i(TAG, "Select by Age");
    uri=Uri.parse("content://"+MatcherConst.AUTHORITY+"/"+DBConst.TABLE_PERSON+
        "/ByAge/"+age);
    selectRecord(uri);

    Log.i(TAG, "Select by Sex");
    uri=Uri.parse("content://"+MatcherConst.AUTHORITY+"/"+DBConst.TABLE_PERSON+
        "/BySex/"+sex);
    selectRecord(uri);

    Log.i(TAG, "Select All");
    uri=Uri.parse("content://"+MatcherConst.AUTHORITY+"/"+DBConst.TABLE_PERSON);
    selectRecord(uri);
}

private void selectRecord(Uri uri){
    Cursor cursor;
    cursor=resolver.query(uri,new String[]{DBConst.COLUMN_AGE, DBConst.COLUMN_NAME,
        DBConst.COLUMN_SEX},null,null,null);
    if(cursor!=null){
```

```
            while(cursor.moveToNext()){
                Log.i(TAG, "name = "+cursor.getString(1)+ " age = "+cursor.getInt(0) +" "+cursor.getString(2));
            }
        }
    }
```

输出结果如下。

```
I/com.breakloop.contentproviderdemo2.MainActivity: Select by Name
I/com.breakloop.contentproviderdemo2.MainActivity: name = A age = 10 Male
I/com.breakloop.contentproviderdemo2.MainActivity: Select by Age
I/com.breakloop.contentproviderdemo2.MainActivity: name = B age = 11 Male
I/com.breakloop.contentproviderdemo2.MainActivity: Select by Sex
I/com.breakloop.contentproviderdemo2.MainActivity: name = A age = 10 Male
I/com.breakloop.contentproviderdemo2.MainActivity: name = B age = 11 Male
I/com.breakloop.contentproviderdemo2.MainActivity: Select All
I/com.breakloop.contentproviderdemo2.MainActivity: name = A age = 10 Male
I/com.breakloop.contentproviderdemo2.MainActivity: name = B age = 11 Male
I/com.breakloop.contentproviderdemo2.MainActivity: name = C age = 12 Female
I/com.breakloop.contentproviderdemo2.MainActivity: name = D age = 13 Female
```

更新代码如下。

```
public void updateRecord(){
    Uri uri;
    String name="A";
    int age=11;
    String sex="Female";
    ContentValues values=new ContentValues();

    Log.i(TAG, "Update by Name");
    uri=Uri.parse("content://"+MatcherConst.AUTHORITY+"/"+DBConst.TABLE_PERSON+
        "/ByName/"+name);
    values.put(DBConst.COLUMN_NAME,name+name);
    update(uri,values);

    Log.i(TAG, "Update by Age");
    uri=Uri.parse("content://"+MatcherConst.AUTHORITY+"/"+DBConst.TABLE_PERSON+
        "/ByAge/"+age);
    values.clear();
    values.put(DBConst.COLUMN_AGE,14);
    update(uri,values);

    Log.i(TAG, "Update by Sex");
    uri=Uri.parse("content://"+MatcherConst.AUTHORITY+"/"+DBConst.TABLE_PERSON+
        "/BySex/"+sex);
    values.clear();
    values.put(DBConst.COLUMN_AGE,15);
    update(uri,values);
```

```
        Log.i(TAG, "Update All");
        uri=Uri.parse("content://"+MatcherConst.AUTHORITY+"/"+DBConst.TABLE_PERSON);
        values.put(DBConst.COLUMN_SEX,"Male");
        update(uri,values);

        Log.i(TAG, "Select All");
        uri=Uri.parse("content://"+MatcherConst.AUTHORITY+"/"+DBConst.TABLE_PERSON);
        selectRecord(uri);
}

private void update(Uri uri,ContentValues values){
        int count;
        count=resolver.update(uri,values,null,null);
        Log.i(TAG, "update "+count+" record");
}
```

更新后的结果如下。

```
I/com.breakloop.contentproviderdemo2.MainActivity: Update by Name
I/com.breakloop.contentproviderdemo2.MainActivity: update 1 record
I/com.breakloop.contentproviderdemo2.MainActivity: Update by Age
I/com.breakloop.contentproviderdemo2.MainActivity: update 1 record
I/com.breakloop.contentproviderdemo2.MainActivity: Update by Sex
I/com.breakloop.contentproviderdemo2.MainActivity: update 2 record
I/com.breakloop.contentproviderdemo2.MainActivity: Update All
I/com.breakloop.contentproviderdemo2.MainActivity: update 4 record
I/com.breakloop.contentproviderdemo2.MainActivity: Select All
I/com.breakloop.contentproviderdemo2.MainActivity: name = AA age = 15 Male
I/com.breakloop.contentproviderdemo2.MainActivity: name = B age = 15 Male
I/com.breakloop.contentproviderdemo2.MainActivity: name = C age = 15 Male
I/com.breakloop.contentproviderdemo2.MainActivity: name = D age = 15 Male
```

删除 4 条记录(这里是对更新前的数据库进行的操作!)

```
public void deleteRecord(){
        Uri uri;
        String name="A";
        int age=11;
        String sex="Female";

        Log.i(TAG, "Delete by Name");
        uri=Uri.parse("content://"+MatcherConst.AUTHORITY+"/"+DBConst.TABLE_PERSON+
            "/ByName/"+name);
        delete(uri);

        Log.i(TAG, "Select All");
        uri=Uri.parse("content://"+MatcherConst.AUTHORITY+"/"+DBConst.TABLE_PERSON);
        selectRecord(uri);
```

```
        Log.i(TAG, "Delete by Age");
        uri=Uri.parse("content://"+MatcherConst.AUTHORITY+"/"+DBConst.TABLE_PERSON+
            "/ByAge/"+age);
        delete(uri);

        Log.i(TAG, "Select All");
        uri=Uri.parse("content://"+MatcherConst.AUTHORITY+"/"+DBConst.TABLE_PERSON);
        selectRecord(uri);

        Log.i(TAG, "Delete by Sex");
        uri=Uri.parse("content://"+MatcherConst.AUTHORITY+"/"+DBConst.TABLE_PERSON+
            "/BySex/"+sex);
        delete(uri);

        Log.i(TAG, "Select All");
        uri=Uri.parse("content://"+MatcherConst.AUTHORITY+"/"+DBConst.TABLE_PERSON);
        selectRecord(uri);

        Log.i(TAG, "Delete All");
        uri=Uri.parse("content://"+MatcherConst.AUTHORITY+"/"+DBConst.TABLE_PERSON);
        delete(uri);

        Log.i(TAG, "Select All");
        uri=Uri.parse("content://"+MatcherConst.AUTHORITY+"/"+DBConst.TABLE_PERSON);
        selectRecord(uri);
    }

    private void delete(Uri uri){
        int count;
        count=resolver.delete(uri,null,null);
        Log.i(TAG, "delete "+count+" record");
    }
```

删除后的结果如下。

```
I/com.breakloop.contentproviderdemo2.MainActivity: delete 1 record
I/com.breakloop.contentproviderdemo2.MainActivity: Select All
I/com.breakloop.contentproviderdemo2.MainActivity: name = B age = 11 Male
I/com.breakloop.contentproviderdemo2.MainActivity: name = C age = 12 Female
I/com.breakloop.contentproviderdemo2.MainActivity: name = D age = 13 Female
I/com.breakloop.contentproviderdemo2.MainActivity: Delete by Age
I/com.breakloop.contentproviderdemo2.MainActivity: delete 1 record
I/com.breakloop.contentproviderdemo2.MainActivity: Select All
I/com.breakloop.contentproviderdemo2.MainActivity: name = C age = 12 Female
I/com.breakloop.contentproviderdemo2.MainActivity: name = D age = 13 Female
I/com.breakloop.contentproviderdemo2.MainActivity: Delete by Sex
I/com.breakloop.contentproviderdemo2.MainActivity: delete 2 record
```

```
I/com.breakloop.contentproviderdemo2.MainActivity: Select All
I/com.breakloop.contentproviderdemo2.MainActivity: Delete All
I/com.breakloop.contentproviderdemo2.MainActivity: delete 0 record
I/com.breakloop.contentproviderdemo2.MainActivity: Select All
1
```

至此，ContentProvider 的创建和使用便介绍完了。

【单元小结】

- ContentProvider 概述。
- 自定义 ContentProvider。
- Uri 和 UriMatcher 的使用。

【单元自测】

1. 内容提供器(ContentProvider)的方式有哪些？
2. 自定义 ContentProvider 的步骤是什么？
3. Android 应用继承(　　)类，并重写该类用于提供数据和存储数据的方法，就可以向其他应用共享其数据。

  A. ContentProvider   B. ProviderContent
  C. Content   D. Provider

4. 要获取 ContentResolver 对象，可以使用 Activity 提供的(　　)方法。

  A. getContentResolver()   B. getContent()
  C. getResolver()   D. getInitialContentResolver()

5. Authorities 属性唯一标识了一个 ContentProvider，还可以通过(　　)和(　　)来设置其操作权限。

  A. setReadPermission()   B. setWritePermission()
  C. setPermission()   D. getPermission()

【上机实战】

上机目标

- 使用 ContentResolver 操作 ContentProvider 中的数据。
- 监听 ContentProvider 中数据的变化。

上机练习

【问题描述】

使用 ContentProvider 共享并自定义 ContentProvider。

【问题分析】

- 如何共享数据。
- URI 如何操作数据。
- 如何监听数据的变化。

【参考步骤】

创建 Android 工程。请参考书中前述步骤。

【拓展作业】

1. 使用 ContentProvider 共享数据。
2. 使用 ContentResolver 操作 ContentProvider 中的数据。

单元十二 网络编程

课程目标

- 网络编程概述
- HTTP 网络编程
- URL 网络编程
- WebView
- 网络框架

 简介

在 Android 系统中，HTTP 网络编程、URL 网络编辑、WebView、网络框架的相关应用也非常广泛，本单元针对这四部分内容做简单介绍。

12.1 网络编程概述

如今，在 Android 的应用程序中使用网络的应用越来越多，如 QQ、微博、新闻等常见的应用都会大量地使用网络技术；传统的本地应用如音乐播放器、词典等，也加入了在线存储、在线推荐、分享等功能。

随着互联网的不断发展，Android 系统的功能已远超过普通通信手机的功能，更像是有手机功能的 PC，其对网络编程的支持也日益强大。Android 网络编程将会变得更加简捷和广泛：一方面 Android 的开源和强大的开发框架大大简化了网络应用的编程；另一方面众多网络服务提供商的开放 API 也对网络编程提供了极大的便利。

本单元主要讲述如何在手机端使用 HTTP(超文本传送协议)和服务器端进行网络交互，学习 Android 的 HTTP 和 URL 网络编程。令人遗憾的是，Android 5.0 以后移除了 HttpClient 的相关 API，为了弥补这种不足，本单元将会介绍 okhttp 网络请求框架。

12.1.1 使用 HTTP

如果要深入分析 HTTP，可能需要花费一本书的篇幅，此处只需稍微了解即可。HTTP 的工作原理非常简单，就是客户端向服务器发出一条 HTTP 请求，服务器收到请求之后会返回一些数据给客户端，然后客户端再对这些数据进行解析和处理。一个浏览器的基本工作原理也就是如此了，例如，用手机浏览器打开一个网页，其实就是浏览器向服务器发起了一条 HTTP 请求，接着服务器分析出浏览器想要访问的网页，于是会把该网页的 HTML 代码进行返回，然后手机浏览器的内核对返回的 HTML 代码进行解析，最终将页面展示出来。

12.1.2 Android 网络程序的功能

1. 通信功能

在 Android 系统上传统应用得到了加强，Android 系统为电话、短信和联系人提供了强大的管理功能和智能云备份的能力，一些应用也为电话提供了归属地查询和垃圾短信过滤等功能。微信和 QQ 成为 Android 最基本的聊天应用，满足用户最基本的聊天需要。借助于网络，不仅可以用文本来传输信息，还可以使用图像、语音、视频等方式交互；可以将通信内容完全保存在服务器上，方便随时查看。只要在联网的情况下就可以使用，不必通过电信运营商，大大降低了通信成本。

2. 分享功能

现在社会，分享已经成为不可或缺的基本功能，很多应用程序都增加了分享的功能以实现其社会化，例如，通过社会化应用，用户可以将自己的想法、心情随时随地发到微博上；看到有趣的事物也可以直接上传到分享网站；有需要帮助的问题，可以使用问答应用提问，等待大家的回答。这些分享的信息，不但能充分利用移动的优势，将用户的信息及时发布到互联网上，分享给好友，还能充分利用终端的硬件优势，如 GPS、摄像头等，多维度地分享自己的生活和见闻。用户在分享的过程中，大量的经验和知识被保存起来，为知识化的互联网提供了更多的素材。

Android 网络应用程序丰富了社会化交流，用户可以及时发布信息、提供帮助、交流看法，分享身边的一切。

3. 娱乐游戏

近年来，Android 系统软硬件的快速发展(软件方面，Android 系统原生支持更多的游戏外设；硬件方面，CUP、内存和屏幕等硬件变得更加强大)，使得游戏和娱乐应用不断发展。Android 应用商店里的游戏娱乐比重不断增加，更多的大型游戏出现在 Android 上。同时，使用 Android 进行在线支付、购物变得更加方便。

在线的游戏、视频、音乐、广播已经成为人们生活的一部分，Android 已变为娱乐游戏的强大载体。

4. 企业应用

Android 作为免费开源的系统，可以大大降低企业应用的成本；而开放的源代码又为企业应用提供了更多稳定性的保证。众多公司也根据不同的应用环境开发了不同应用，如汽运、运动、网购等不同方面的 Android 应用产品。

Android 企业应用将进一步提高企业的生产效率，为企业移动办公提供更加方便、可靠的平台。

12.2 HTTP 网络编程

HttpClient 是 Apache 开源组织提供的 HTTP 网络访问接口(一个开源的项目)，从名字上就可以看出，它是一个简单的 HTTP 客户端(并不是浏览器)，可以发送 HTTP 请求，接受 HTTP 响应。但是不会缓存服务器的响应，也不能执行 HTTP 页面中签入、嵌入的 JS 代码，自然也不会对页面内容进行任何解析、处理，这些都是需要开发人员来完成的。

然而 HttpClient 只是一个接口，因此无法创建它的实例，但通常情况下会创建一个 DefaultHttpClient 的实例。HttpClient 有两种请求方式，一种是使用 GET 请求；另一种是使用 POST 请求。

1. HttpGet

在 HttpClient 中，我们可以非常轻松地使用 HttpGet 对象来通过 GET 方式进行数据请求操作，当获得 HttpGet 对象后可以使用 HttpClient 的 execute()方法来向服务器发送请求。在发送的 GET 请求被服务器响应后，会返回一个 HttpResponse 响应对象，利用这个响应对象我们能够获得响应回来的状态码，如 200、400、401 等。

```java
public String HttpGetMethod()
{
    String result = "";
    try
    {
        HttpGet httpRequest = new HttpGet(urlStr);
        HttpClient httpClient = new DefaultHttpClient();
        HttpResponse httpResponse = httpClient.execute(httpRequest);
        if(httpResponse.getStatusLine().getStatusCode()==HttpStatus.SC_OK)
        {
            result = EntityUtils.toString(httpResponse.getEntity());
        }
        else
        {
            result = "null";
        }
        return result;
    }
    catch(Exception e)
    {
        return null;
    }
}
```

2. HttpPost

当我们使用 POST 方式时，可以使用 HttpPost 类来进行操作。当获取了 HttpPost 对象后，就需要向这个请求体传入键值对，该键值对可以使用 NameValuePair 对象进行构造，然后再使用 HttpRequest 对象最终构造请求体，最后使用 HttpClient 的 execute()方法来发送请求，并在得到响应后返回一个 HttpResponse 对象。其他操作与在 HttpGet 对象中的操作一样。

```java
public String HttpPostMethod(String key,String value)
{
    String result = "";
    try
    {
        // HttpPost 连接对象
        HttpPost httpRequest = new HttpPost(urlStr);
        // 使用 NameValuePair 保存要传递的 Post 参数
        List<NameValuePair> params = new ArrayList<NameValuePair>();
        // 添加要传递的参数
        params.add(new BasicNameValuePair(key, value));
```

```
            // 设置字符集
            HttpEntity httpentity = new UrlEncodedFormEntity(params, "UTF-8");
            // 请求 httpRequest
            httpRequest.setEntity(httpentity);
            // 取得默认的 HttpClient
            HttpClient httpclient = new DefaultHttpClient();
            // 取得 HttpResponse
            HttpResponse httpResponse = httpclient.execute(httpRequest);
            // HttpStatus.SC_OK 表示连接成功
            if (httpResponse.getStatusLine().getStatusCode() == HttpStatus.SC_OK) {
                // 取得返回的字符串
                result = EntityUtils.toString(httpResponse.getEntity());
                return result;
            } else {
                return "null";
            }
        }
    catch(Exception e)
    {
        return null;
    }
}
```

总结：如果 Android 不与网络资源进行交互，则它与当初的普通系统没有任何区别，所以网络编程对 Android 开发来说有非常特殊的意义。

12.3 URL 网络编程

HttpURLConnection 是继承自 URLConnection 的一个抽象类，在 HTTP 编程时，来自 HttpURLConnection 的类是所有操作的基础，获取该对象的代码如下。

```
public HttpURLConnection urlconn= null;
private void Init() throws IOException
{
    if (urlStr=="")
    {
        urlStr="http://www.baidu.com";
    }
    URL url = new URL(urlStr);
    //打开一个 URL 所指向的 Connection 对象
    urlconn = (HttpURLConnection)url.openConnection();
}
```

HttpURLConnection 对网络资源的请求在默认情况下是使用 GET 方式的，所以当使用 GET 方式时，不需要我们做太多的工作。

```java
public HttpURLConnection urlconn= null;
private void Init() throws IOException
{
    if (urlStr=="")
    {
        urlStr="http://www.sina.com.cn";
    }
    URL url = new URL(urlStr);
    //打开一个 URL 所指向的 Connection 对象
    urlconn = (HttpURLConnection)url.openConnection();
}
/**
* HTTP 中的 GET 请求，在 URL 中带有请求的参数，请求的 URL 格式通常为
 "http://XXX.XXXX.com/xx.jsp?param=value"
* 在 Android 中默认的 HTTP 请求为 GET 方式
* @return
* @throws IOException
*/
public String HttpGetMethod() throws IOException
{
    if(urlconn == null)
    {
        Init();
    }
    String result = StreamDeal(urlconn.getInputStream());
    urlconn.disconnect();
    return result;
}
```

当我们需要使用 POST 方式时，则要使用 setRequestMethod()方法来设置请求方式。

```java
/**
* HTTP 中的 POST 请求，不在 URL 中附加任何参数，这些参数都会通过 cookie 或 session 等其他
 方式以键值对的形式(key=value)传送到服务器上，完成一次请求
* 请求的 URL 格式通常为:"http://XXX.XXXX.com/xx.jsp"
* @param param  请求的键名
* @param value  请求的数据值
* @throws IOException
*/
public String HttpPostMethod(String key,String value) throws IOException
{
    if (urlconn==null)
    {
        Init();
    }
    //设置该 URLConnection 可读
    urlconn.setDoInput(true);
```

```
//设置该 URLConnection 可写
urlconn.setDoOutput(true);
//使用 POST 方式来提交数据
urlconn.setRequestMethod("POST");
//不运行缓存
urlconn.setUseCaches(false);
//当使用 POST 方式进行数据请求时，我们可以手动执行 connect 动作，当然，这个动作其实
在 getOutputStream()方法中会默认执行
//上面设置 URLConnection 属性的动作，一定要在 connect 动作执行前，因为一旦动作已
//经执行，熟悉设置就没有任何作用了
urlconn.connect();
//使用 POST 方式时，需要自己构造部分 Http 请求的内容，因此我们需要使用
//OutputStream 来进行数据写操作
OutputStreamWriter writer = new OutputStreamWriter(urlconn.getOutputStream());
String urlQueryStr = key+"="+URLEncoder.encode(value, "UTF-8");
writer.write(urlQueryStr);
writer.flush();
writer.close();
//获取返回的内容
String result = StreamDeal(urlconn.getInputStream());
return result;

}
```

12.4 WebView

12.4.1 WebView 介绍

WebView 是 Android 中的一个原生 UI 控件，它是一个基于 webkit 引擎、展现 Web 页面的控件。WebView 在不同高版本之间采用了不同的 webkit 版本内核，4.4 版后直接使用了 Chrome。现在很多 APP 都内置了 Web 网页，如淘宝、京东、聚划算等电商平台。WebView 比较灵活，不需要升级客户端，只需要修改网页代码。

12.4.2 WebView 的作用

WebView 的功能非常强大，有以下几种作用。
- 使用 WebView 浏览网页。它可以加载远程网页的 URL 及本地 assets 资源中的 HTML 网页文件，并且 WebView 还能被当作 Java 代码和 JS 代码之间的桥梁，实现两者之间的交互功能。

- 使用 WebView 加载 HTML 代码。Android 其他控件并不能解析 HTML 标签，而 WebView 提供了 loadData()方法，该方法可以对 HTML 标签进行解析，从而将 HTML 页面正确显示出来。
- 使用 WebView 中的 JavaScript 方法可以调用 Android 的方法，从而实现 HTML 页面和原生 Android 之间的交互。

12.4.3　WebView 的使用

WebView 的用法也非常简单，下面我们就通过一个例子来学习一下。新建一个 WebViewTest 项目，然后修改 activity_main.xml 中的代码，如下所示。

```xml
<LinearLayout xmlns:android="http://schemas.android.com/apk/res/android"
    android:layout_width="match_parent"
    android:layout_height="match_parent" >
    <WebView
        android:id="@+id/web_view"
        android:layout_width="match_parent"
        android:layout_height="match_parent" />
</LinearLayout>
```

可以看到，我们在布局文件中使用到了一个新的控件 WebView，该控件是用来显示网页的，这里的写法很简单，给它设置了一个 ID，并让它充满整个屏幕。然后修改 MainActivity 中的代码，如下所示。

```java
public class MainActivity extends Activity {
    private WebView webView;
    @Override
    protected void onCreate(Bundle savedInstanceState) {
        super.onCreate(savedInstanceState);
        setContentView(R.layout.activity_main);
        webView = (WebView) findViewById(R.id.web_view);
        //设置 WebView 支持 JavaScript
        webView.getSettings().setJavaScriptEnabled(true);
        webView.setWebViewClient(new WebViewClient() {
            @Override
            public boolean shouldOverrideUrlLoading(WebView view, String url) {
                // 根据传入的参数再去加载新的网页
                view.loadUrl(url);
                // 表示当前 WebView 可以处理打开新网页的请求，不用借助系统浏览器
                return true;
            }
        });
        webView.loadUrl("http://www.baidu.com");
    }
}
```

MainActivity 中的代码也很短，首先使用 findViewById()方法获取到 WebView 的实例，然后调用 WebView 的 getSettings()方法设置一些浏览器的属性，这里我们并不去设置过多的属性，只调用了 setJavaScriptEnabled()方法来让 WebView 支持 JavaScript 脚本。

接下来的是非常重要的一部分，即调用 WebView 的 setWebViewClient()方法，并传入了 WebViewClient 的匿名类作为参数，然后重写 shouldOverrideUrlLoading()方法。这表明当需要从一个网页跳转到另一个网页时，我们希望目标网页仍然在当前 WebView 中显示，而不是打开系统浏览器。

最后调用 WebView 的 loadUrl()方法，并将网址传入，即可展示相应网页的内容，在这里我们就可以看到百度的首页。

另外还需要注意，由于本程序使用到了网络功能，而访问网络是需要声明权限的，因此我们还得修改 AndroidManifest.xml 文件，并加入权限声明，如下所示。

```
<manifest xmlns:android="http://schemas.android.com/apk/res/android"
    package="com.hp.test">
    <uses-permission android:name="android.permission.INTERNET" />
    ……
</manifest>
```

12.4.4 WebView 与 H5 混合开发

在 Android 开发中，越来越多的商业项目使用了原生 Android 与 H5 进行混合开发，当然不仅是显示一个 WebView 那么简单，有时候还需要本地 Java 代码与 H5 中的 JavaScript 进行交互，Android 也对交互做了很好的封装，所以很容易实现。例如：单击网页中的按钮调用 Android 原生对话框，单击网页中的电话号码调用 Android 拨号 APP 等。

下面给大家介绍一下如何实现 Android 与 H5 的交互。

1. Android 调用 H5 中无参函数

有时我们需要单击 Android 的按钮改变 HTML 页面的内容，当单击"显示"按钮时，显示 HTML 中的一段内容，如图 12-1 和图 12-2 所示。

具体实现的代码如下。

(1) 在 AndroidManifest.xml 文件中添加网络权限。

```
<uses-permission android:name="android.permission.INTERNET"/>
```

(2) 在 app 目录下创建 assets 文件夹，然后在 assets 文件夹下添加要使用的 HTML 文件 test.html。test.html 的代码如下。

```
<html>
    <head>
        <!--从 assets 中加载的中文网页会出现乱码，解决办法是给 HTML 指定编码-->
        <meta http-equiv="content-type" content="text/html;charset=UTF-8">
        <title> </title>
```

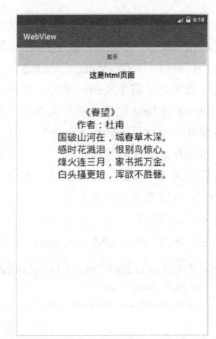

图 12-1　　　　　　　　　　　　　图 12-2

```
<script type="text/javascript">
    //无参无返回值的函数
    function show(){
        //显示诗--春望
        document.getElementById("p").style.display="inline";
    }
</script>
</head>
<body>
    <h3 align="center">这是 html 页面</h3> <br/>
    <pre id="p" style="display: none;"><font color="blue"    size=5    face="隶书" >
        《春望》
        作者：杜甫
国破山河在，城春草木深。
感时花溅泪，恨别鸟惊心。
烽火连三月，家书抵万金。
白头搔更短，浑欲不胜簪。
        </font>
    </pre>
</body>
</html>
```

以上代码包括 3 个基本点：①在 head 的 meta 标签中设置编码，用于保证加载的网页不会出现乱码；②show()为 JavaScript 的无参函数，用于 Android 原生控件调用；③body 内部为 html 的主体部分，用于 WebView 的显示。

(3) 在 layout 文件夹添加 activity_webview.xml。

```xml
<LinearLayout
    xmlns:android="http://schemas.android.com/apk/res/android"
    xmlns:app="http://schemas.android.com/apk/res-auto"
    xmlns:tools="http://schemas.android.com/tools"
    android:layout_width="match_parent"
    android:layout_height="match_parent"
    android:orientation="vertical"
    tools:context="com.hp.ui.WebViewActivity">

    <Button
        android:id="@+id/button"
        android:layout_width="match_parent"
        android:layout_height="wrap_content"
        android:text="显示"/>
    <WebView
        android:id="@+id/webView"
        android:layout_width="match_parent"
        android:layout_height="match_parent" />

</LinearLayout>
```

(4) 在 MainActivity.java 中的具体引用与实现。

```java
import android.os.Bundle;
import android.support.v7.app.AppCompatActivity;
import android.view.View;
import android.webkit.WebSettings;
import android.webkit.WebView;
import android.widget.Button;

import com.hp.main.R;

public class MainActivity    extends AppCompatActivity {

    private WebView webView;
    private Button button;
    @Override
    protected void onCreate(Bundle savedInstanceState) {
        super.onCreate(savedInstanceState);
        setContentView(R.layout.activity_webview);
        button = (Button)findViewById(R.id.button);
        webView = (WebView)findViewById(R.id.webView);
        WebSettings webSettings = webView.getSettings();
        webSettings.setJavaScriptEnabled(true);
        webView.loadUrl("file:///android_asset/test.html");
        button.setOnClickListener(new View.OnClickListener() {
            @Override
```

```
                    public void onClick(View v) {
                        //直接访问 H5 中不带参数的方法，show()为 H5 中的方法
                        webView.loadUrl("JavaScript:show()");
                    }
                });
        }
    }
```

以上代码中，WebView 是加载 HTML 文件的控件，button 是要单击的控件，WebSettings 用于管理 WebView 状态的对象，webSettings.setJavaScriptEnabled(true)表示设置为可调用 js 方法，webView.loadUrl("file:///android_asset/test.html")是 WebView 加载本地 HTML 文件的代码，webView.loadUrl("JavaScript:show()")表示当单击 button 按钮时调用 js 函数的代码。

2. Android 调用 H5 中有参函数

有时我们需要将 Android 的数据传到 HTML，HTML 接收数据后再做进一步的操作，假如我们单击 Android 的按钮时传入一个字符串，然后 HTML 通过弹出框将传入的字符串提示出来，如图 12-3 和图 12-4 所示。

图 12-3　　　　　　　　　　图 12-4

具体实现的代码如下。

(1) 在 AndroidManifest.xml 文件中添加网络权限。

```
<uses-permission android:name="android.permission.INTERNET"/>
```

(2) 在 app 目录下创建 assets 文件夹，然后在 assets 文件夹下添加要使用的 HTML 文件 test.html，其中 test.html 的代码如下。

```
<html>
```

```html
<head>
<!--从 assets 中加载的中文网页会出现乱码，解决办法是给 html 指定编码-->
<meta http-equiv="content-type" content="text/html;charset=utf-8">
<title> </title>
<script type="text/javascript">
    //有参无返回值的函数
    function alertMessage(message) {
        alert(message);
    }
</script>
</head>
<body>
<h3 align="center">这是 html 页面</h3> <br/>

</body>
</html>
```

以上代码包括 3 个基本点：①在 head 的 meta 标签中设置编码，用于保证加载的网页不会出现乱码；②alertMessage(message)为 JavaScript 的带参数的函数，用于 Android 原生控件调用；③body 内部为 html 的主体部分，用于 WebView 的显示。

(3) 在 layout 文件夹添加 activity_webview.xml。

```xml
<LinearLayout
    xmlns:android="http://schemas.android.com/apk/res/android"
    xmlns:app="http://schemas.android.com/apk/res-auto"
    xmlns:tools="http://schemas.android.com/tools"
    android:layout_width="match_parent"
    android:layout_height="match_parent"
    android:orientation="vertical"
    tools:context="com.hp.ui.MainActivity">

    <Button
        android:id="@+id/button"
        android:layout_width="match_parent"
        android:layout_height="wrap_content"
        android:text="显示"/>
    <WebView
        android:id="@+id/webView"
        android:layout_width="match_parent"
        android:layout_height="match_parent" />

</LinearLayout>
```

(4) 在 MainActivity.java 中的具体引用与实现。

```java
import android.os.Bundle;
import android.support.v7.app.AppCompatActivity;
import android.view.View;
```

```
import android.webkit.WebChromeClient;
import android.webkit.WebSettings;
import android.webkit.WebView;
import android.webkit.WebViewClient;
import android.widget.Button;

import com.hp.main.R;

public class MainActivity extends AppCompatActivity {

    private WebView webView;
    private Button button;
    @Override
    protected void onCreate(Bundle savedInstanceState) {
        super.onCreate(savedInstanceState);
        setContentView(R.layout.activity_webview);
        button = (Button)findViewById(R.id.button);
        webView = (WebView)findViewById(R.id.webView);
        WebSettings webSettings = webView.getSettings();
        webSettings.setJavaScriptEnabled(true);

        webView.loadUrl("file:///android_asset/test.html");
        button.setOnClickListener(new View.OnClickListener() {
            @Override
            public void onClick(View v) {

                //直接访问 H5 中带参数的方法，alertMessage()为 H5 中的方法
                webView.loadUrl("JavaScript:alertMessage('调用了带参的 js 方法')");
            }
        });
    }
}
```

以上代码中，WebView 是加载 HTML 文件的控件，button 是要单击的控件，WebSettings 用于管理 WebView 状态的对象，webSettings.setJavaScriptEnabled(true)表示设置为可调用 js 方法，webView.loadUrl("file:///android_asset/test.html")是 WebView 加载本地 HTML 文件的代码，webView.loadUrl("JavaScript:alertMessage('调用了带参的 js 方法')")则是单击 button 按钮时调用 js 函数的代码，注意 js 函数传入的字符串使用的是单元号。

3. H5 调用 Android 中无参方法

有时我们在进行混合开发时不仅需要 Android 调用 H5 函数，也需要 H5 调用 Android 的原生方法。那么怎么调用呢？接下来看下面的例子，这里我们实现单击 H5 页面的按钮，调用 Android 原生的方法弹出 Toast 提示，如图 12-5 和图 12-6 所示。

图 12-5　　　　　　　　图 12-6

具体实现的代码如下。

(1) 在 layout 文件夹添加 activity_webview.xml。

```
<LinearLayout
    xmlns:android="http://schemas.android.com/apk/res/android"
    xmlns:app="http://schemas.android.com/apk/res-auto"
    xmlns:tools="http://schemas.android.com/tools"
    android:layout_width="match_parent"
    android:layout_height="match_parent"
    android:orientation="vertical"
    tools:context="com.hp.ui.MainActivity">
    <WebView
        android:id="@+id/webView"
        android:layout_width="match_parent"
        android:layout_height="match_parent" />

</LinearLayout>
```

(2) 在 MainActivity.java 中实现以下代码。

```
import android.os.Bundle;
import android.support.v7.app.AppCompatActivity;
import android.webkit.JavascriptInterface;
import android.webkit.WebChromeClient;
import android.webkit.WebSettings;
import android.webkit.WebView;
import android.webkit.WebViewClient;
import android.widget.Toast;

import com.hp.main.R;
```

```java
public class MainActivity extends AppCompatActivity {

    private WebView webView;
    @Override
    protected void onCreate(Bundle savedInstanceState) {
        super.onCreate(savedInstanceState);
        setContentView(R.layout.activity_webview);
        webView = (WebView)findViewById(R.id.webView);
        WebSettings webSettings = webView.getSettings();
        webSettings.setJavaScriptEnabled(true);
        webView.loadUrl("file:///android_asset/test3.html");
        webView.setWebViewClient(new WebViewClient());
        webView.setWebChromeClient(new WebChromeClient());
        /** 打开 js 接口给 H5 调用，参数 1 为本地类名，参数 2 为别名；
         * H5 用 window.别名.类名中的方法名才能调用方法里的内容，
         * 例如：window.android.show();
         */
        webView.addJavascriptInterface(new JsInteration(), "android");
    }

    /*** 自己写一个类，里面是提供给 H5 访问的方法 **/
    public class JsInteration {

        //一定要写，否则 H5 调不到这个方法
        @JavascriptInterface
        public void show() {
            Toast.makeTextMainActivity.this,"H5 调用了此方法",Toast.LENGTH_LONG).show();
        }
    }

}
```

以上代码中，WebView 是加载 HTML 文件的控件，button 是要单击的控件，WebSettings 用于管理 WebView 状态的对象，webSettings.setJavaScriptEnabled(true)表示设置为可调用 js 方法，WebView.loadUrl("file:///android_asset/test.html")是 WebView 加载本地 HTML 文件的代码。这里 WebView 添加了 setWebViewClient()和 setWebChromeClient()两个方法，是为了更好地兼容 JavaScript 的调用。

此处定义一个内部类 JsInteration，类中定义一个给 H5 调用的 show()方法，该方法一定要加上@JavascriptInterface，否则 H5 调不到该方法，然后使用 addJavascriptInterface(new JsInteration(), "android") 打开 js 接口给 H5 调用，其中第一个参数为本地类名，第二个参数为别名；H5 用 window.别名.类名中的方法名才能调用方法里的内容，如 window.android.show()。

(3) 在 app 目录下创建 assets 文件夹，然后在 assets 文件夹下添加要使用的 HTML 文件 test.html，其中 test.html 的代码如下。

```html
<html>
    <head>
        <!--从 assets 中加载的中文网页会出现乱码，解决办法是给 html 指定编码-->
        <meta http-equiv="content-type" content="text/html;charset=UTF-8">
        <title> </title>
        <script type="text/javascript">

            function showToast(){
                //调用原生的方法，android 为约定的别名；show()为原生的方法
                window.android.show();
            }
        </script>
    </head>
    <body>
    <h3 align="center">这是 html 页面</h3> <br/>
        <button onclick="showToast()">单击调用本地方法</button>

    </body>
</html>
```

以上代码中，showToast()为 js 调用 Android 原生方法的函数，通过 window.android.show() 来调用，其中 Android 为约定别名，show()为原生方法，当单击 HTML 中的 button 按钮时调用该方法。

4. H5 调用 Android 中有参方法

有时我们还需要 H5 向 Android 传递参数。那么怎么调用呢？接下来看下面的例子，这里我们实现 H5 登录页面，通过 Android 获取登录信息实现登录，如图 12-7 和图 12-8 所示。

图 12-7   图 12-8

具体实现的代码如下。

(1) 在 layout 文件夹添加 activity_webview.xml。

```xml
<LinearLayout
    xmlns:android="http://schemas.android.com/apk/res/android"
    xmlns:app="http://schemas.android.com/apk/res-auto"
    xmlns:tools="http://schemas.android.com/tools"
    android:layout_width="match_parent"
    android:layout_height="match_parent"
    android:orientation="vertical"
    tools:context="com.hp.ui.MainActivity">
    <WebView
        android:id="@+id/webView"
        android:layout_width="match_parent"
        android:layout_height="match_parent" />

</LinearLayout>
```

(2) 在 MainActivity.java 中实现以下代码。

```java
import android.os.Bundle;
import android.support.v7.app.AppCompatActivity;
import android.webkit.JavascriptInterface;
import android.webkit.WebChromeClient;
import android.webkit.WebSettings;
import android.webkit.WebView;
import android.webkit.WebViewClient;
import android.widget.Toast;

import com.hp.main.R;

public class MainActivity extends AppCompatActivity {

    private WebView webView;
    @Override
    protected void onCreate(Bundle savedInstanceState) {
        super.onCreate(savedInstanceState);
        setContentView(R.layout.activity_webvie);
        webView = (WebView)findViewById(R.id.webView);
        WebSettings webSettings = webView.getSettings();
        webSettings.setJavaScriptEnabled(true);
        webView.loadUrl("file:///android_asset/test4.html");
        webView.setWebViewClient(new WebViewClient());
        webView.setWebChromeClient(new WebChromeClient());
        /** 打开 js 接口给 H5 调用，参数 1 为本地类名，参数 2 为别名；
         * H5 用 window.别名.类名中的方法名才能调用方法里的内容，
         * 例如：window.android.show();
         */
```

```java
            webView.addJavascriptInterface(new JsInteration(), "android");
        }

        /*** 自己写一个类，里面是提供给 H5 访问的方法 **/
        public class JsInteration {

            //一定要写，否则 H5 调不到这个方法
            @JavascriptInterface
            public void login(String username, String password) {
                Toast.makeText(MainActivity.this,"用户名："+username+", 密码：
"+password,Toast.LENGTH_LONG).show();
            }
        }
```

以上代码中，WebView 是加载 HTML 文件的控件，button 是要单击的控件，WebSettings 用于管理 WebView 状态的对象，webSettings.setJavaScriptEnabled(true)表示设置为可调用 js 方法，WebView.loadUrl("file:///android_asset/test.html")是 WebView 加载本地 HTML 文件的代码。这里 WebView 添加了 setWebViewClient()和 setWebChromeClient()两个方法，是为了更好地兼容 JavaScript 的调用。

这里定义一个内部类 JsInteration，类中定义一个给 H5 调用的 show()方法，该方法一定要加上@JavascriptInterface，否则 H5 调不到该方法，然后使用 addJavascriptInterface(new JsInteration(), "android") 打开 js 接口给 H5 调用，其中第一个参数为本地类名，第二个参数为别名；H5 用 window.别名.类名中的方法名才能调用方法里的内容，login()方法为 Android 登录方法，在这里我们获取用户名和密码后提示出来。

（3）在 app 目录下创建 assets 文件夹，然后在 assets 文件夹下添加要使用的 HTML 文件 test.html，其中 test.html 的代码如下。

```html
<html>
    <head>
        <!--从 assets 中加载的中文网页会出现乱码，解决办法是给 html 指定编码-->
        <meta http-equiv="content-type" content="text/html;charset=UTF-8">
        <title> </title>
        <style type="text/css">
            table{
                margin:20px auto;}
        </style>
        <script type="text/javascript">

            function login(){
                var username = document.getElementById("username").value;
                var password = document.getElementById("password").value;
                //调用原生的方法，android 为约定的别名；login()为原生的方法
                window.android.login(username,password);
            }
        </script>
```

```
            </head>
            <body>
                <h3 align="center">html 登录页面</h3>
                    <form>
                        <table>
                            <Tr>
                                <td>用户名：</td>
                                <td><input type="text" id="username"/></td>
                            </Tr>
                            <Tr>
                                <td>密码：</td>
                                <td><input type="text" id="password"/></td>
                            </Tr>
                            <Tr><td colspan="2" align="center">
                                <input value="登录" type="button" onClick="login()"/>
                            </td></Tr>
                        </table>
                    </form>
            </body>
        </html>
}
```

以上代码中，login()为js调用Android原生方法的函数，我们通过标签的id拿到输入框的值后，通过window.android.login((username,password)调用，将获取的用户名和密码传给Android。

12.5 网络框架

12.5.1 网络框架简介

Android 程序最重要的模块就是网络部分，如何从网络上下载数据、将处理过的数据上传至网络，往往是Android 程序的关键环节。

Android 原生提供基于 HttpClient 和 HttpUrlConnection 两种网络访问方式。利用原生的这两种方式编写网络代码，需要自己考虑很多，获取数据或许可以，但是如果要将手机本地数据上传至网络就造成了很大的工作量。在 Android 5.0 版本中，Google 就不推荐使用 HttpClient 了，到了 Android 6.0 (api 23) SDK 版本，不再提供 org.apache.http.* (只保留几个类)，因此，设置 android SDK 的编译版本为 23 时，使用了 httpClient 相关类的库项目，如 android-async-http 等，会出现有一些类找不到的错误。

okhttp 作为一个处理网络请求的开源项目，是 Android 当前最热门的网络框架，由移动支付 Square 公司贡献，用于替代 HttpUrlConnection 和 Apache HttpClient(android API23 6.0 中已移除 HttpClient)。所以 okhttp 是目前比较适合的网络框架，根据业务需求进行适当的封装能够很好地适用于我们的需求。另外，该框架现在已经被广泛使用，如著名的公司 Facebook，他们在自己的 Android 客户端中对网络访问使用的就是 okhttp。现在 Google 官方也从 Android 4.4 开始使用 okhttp 作为 HttpURLConnection 的默认实现了。okhttp 支持在糟糕的网络环境下面更快地重试，并且还能利用 SPDY 协议进行快速的并发网络请求。

12.5.2　okhttp 框架的优点和作用

okhttp 作为一款优秀的开源网络框架，主要有以下几个优点。
- 支持 HTTP2/SPDY(SPDY 是 Google 开发的基于 TCP 的传输层协议，用以最小化网络延迟，提升网络速度，优化用户的网络使用体验。)
- socket 自动选择最好路线，并支持自动重连，拥有自动维护的 socket 连接池，减少握手次数，减少了请求延迟，共享 Socket，减少对服务器的请求次数。
- 基于 Headers 的缓存策略减少重复的网络请求。
- 拥有 Interceptors 轻松处理请求与响应(自动处理 GZip 压缩)。

okhttp 有以下几种作用。
- okhttp 支持 PUT、DELETE、POST、GET 等请求方法,可以像 PC 端一样设置 HTTP 请求的请求头，数据不仅可以以键值对的形式传输，也可以封装成 JSON 字符串传输。
- 文件的上传下载，实现通过文件路径自动封装上传，不再需要先将文件转换成流后再上传，下载实现断点续传的功能。
- 在加载图片时 okhttp 内部会对图片大小自动压缩，防止手机内存溢出。
- okhttp 不仅支持请求回调，而且还可以直接返回对象、对象集合。
- 支持 session 的保持，手机端可以通过 okhttp 访问服务器保存的用户数据，很好地解决了手机端和服务端不能维持状态的问题。

12.5.3　okhttp 的使用

使用 okhttp 进行网络请求主要有 POST 和 GET 两种请求方式，而它的网络请求可以是同步请求也可以是异步请求。下面分情况进行介绍。

我们需要先引入 okhttp 的包，这就需要在项目的 app 目录下的 build.gradle 文件下 okhttp 和 okio 两个引入。

```
apply plugin: 'com.android.application'
..........
dependencies {
    compile fileTree(dir: 'libs', include: ['*.jar'])
    testCompile 'junit:junit:4.12'
    compile 'com.android.support:appcompat-v7:25.3.1'
    compile 'com.squareup.okhttp:okhttp:2.7.5'
    compile 'com.squareup.okio:okio:1.7.0'
```

```
            compile 'com.android.support:design:25.3.1'
}
```

1. GET 的同步请求

我们知道 Android 网络请求需要放在子线程里进行，所以对于同步请求在请求时需要开启子线程，请求成功后需要跳转到 UI 线程修改 UI。

首先本程序使用到了网络功能，而访问网络是需要声明权限的，因此我们还得修改 AndroidManifest.xml 文件，并加入权限声明，如下所示。

```xml
<manifest xmlns:android="http://schemas.android.com/apk/res/android"
    package="com.hp.test">
    <uses-permission android:name="android.permission.INTERNET" />
    ……
</manifest>
```

我们要在 MainActivity.java 中实现以下代码。

```java
import android.os.Bundle;
import android.support.v7.app.AppCompatActivity;
import android.util.Log;

import com.hp.main.R;
import com.squareup.okhttp.Call;
import com.squareup.okhttp.okhttpClient;
import com.squareup.okhttp.Request;
import com.squareup.okhttp.Response;

public class MainActivity extends AppCompatActivity {
    @Override
    protected void onCreate(Bundle savedInstanceState) {
        super.onCreate(savedInstanceState);
        setContentView(R.layout.activity_main);
        getDatasync();
    }
    public void getDatasync(){
        new Thread(new Runnable() {
            @Override
            public void run() {
                try {
                    //创建 okhttpClient 对象
                    okhttpClient client = new okhttpClient();
                    Request request = new Request.Builder()
                            .url("http://www.baidu.com")//请求接口
                            .build();//创建 Request 对象
                    Call call = client.newCall(request);
                    //得到 Response 对象
                    Response response = call.execute();
```

```
                    if (response.isSuccessful()) {
                        Log.d("MainActivity","response.code()=="+response.code());
                        Log.d("MainActivity","response.message()=="+response.message());
                    }
                } catch (Exception e) {
                    e.printStackTrace();
                }
            }
        }).start();
    }
}
```

此时运行会打印出如图 12-9 所示的结果。

```
/MainActivity: response.code()==200
/MainActivity: response.message()==OK
```

图 12-9

Response.code 是 http 响应行中的 code，如果访问成功，则返回 200，说明请求成功，这里我们先创建一个 okhttpClient 对象，接着创建 Request 的 Builder 对象并设置 URL 请求接口,最后通过 build()方法创建 Request。然后通过 okhttpClient 的 newCall(request) 方法实例化 Call 对象，再通过 call.execute()同步请求的方法得到 Response 对象的结果集。response.isSuccessful()是判断我们此次请求是否成功的方法，如果成功可以进行下一步操作。

2. GET 的异步请求

我们既然使用了网络请求框架，那么就要开启子线程，这样很麻烦，有没有更简单的办法呢？其实 okhttp 已经对线程进行了封装,我们接下来看怎么用异步的 get 请求实现登录功能，如图 12-10 所示。

图 12-10

首先本程序使用到了网络功能，而访问网络是需要声明权限的，因此我们还得修改 AndroidManifest.xml 文件，并加入权限声明，如下所示。

```
<manifest xmlns:android="http://schemas.android.com/apk/res/android"
    package="com.hp.test">
    <uses-permission android:name="android.permission.INTERNET" />
    ……
</manifest>
```

布局文件 activity_main.xml 的代码如下。

```
<LinearLayout xmlns:android="http://schemas.android.com/apk/res/android"
```

```xml
xmlns:app="http://schemas.android.com/apk/res-auto"
xmlns:tools="http://schemas.android.com/tools"
android:layout_width="match_parent"
android:layout_height="match_parent"
android:orientation="vertical"
tools:context="com.hp.okhttp.okhttpActivity1">
<LinearLayout
    android:layout_width="match_parent"
    android:layout_height="wrap_content"
    android:orientation="horizontal">
    <TextView
        android:layout_width="wrap_content"
        android:layout_height="wrap_content"
        android:text="用户名:"/>
    <EditText
        android:id="@+id/username"
        android:layout_width="match_parent"
        android:layout_height="wrap_content" />
</LinearLayout>
<LinearLayout
    android:layout_width="match_parent"
    android:layout_height="wrap_content"
    android:orientation="horizontal">
    <TextView
        android:layout_width="wrap_content"
        android:layout_height="wrap_content"
        android:text="密码:"/>
    <EditText
        android:id="@+id/password"
        android:layout_width="match_parent"
        android:layout_height="wrap_content" />
</LinearLayout>
<Button
    android:id="@+id/loginBt"
    android:layout_width="match_parent"
    android:layout_height="wrap_content"
    android:text="登录"
    android:layout_margin="10dip"/>
</LinearLayout>
```

我们要在 MainActivity.java 中实现以下代码。

```java
import android.os.Bundle;
import android.support.v7.app.AppCompatActivity;
import android.util.Log;
import android.view.View;
import android.widget.Button;
import android.widget.EditText;
```

```java
import com.hp.main.R;
import com.squareup.okhttp.Call;
import com.squareup.okhttp.Callback;
import com.squareup.okhttp.okhttpClient;
import com.squareup.okhttp.Request;
import com.squareup.okhttp.Response;

import java.io.IOException;

public class MainActivity extends AppCompatActivity {
    private EditText username;
    private EditText password;
    private Button loginBt;
    @Override
    protected void onCreate(Bundle savedInstanceState) {
        super.onCreate(savedInstanceState);
        setContentView(R.layout.activity_main);
        username = (EditText)findViewById(R.id.username);
        password = (EditText)findViewById(R.id.password);
        loginBt = (Button)findViewById(R.id.loginBt);
        loginBt.setOnClickListener(new View.OnClickListener() {
            @Override
            public void onClick(View v) {
                login();
            }
        });
    }
    public void login(){
        String uname = username.getText().toString();
        String pwd = password.getText().toString();
        String url = "http://172.16.63.134:8080/apms/app/login?username="+uname+"&password="+pwd;
        okhttpClient client = new okhttpClient();
        Request request = new Request.Builder()
                .url(url)
                .build();
        Call call = client.newCall(request);
        call.enqueue(new Callback() {
            @Override
            public void onFailure(Request request, IOException e) {
            }
            @Override
            public void onResponse(Response response) throws IOException {
                if(response.isSuccessful()){
                    //回调的方法执行在子线程
                    Log.d("MainActivity","登录成功了");
```

```
                    Log.d("MainActivity","response.code()=="+response.code());
                    Log.d("MainActivity","response.body().string()=="+response.body().string());
                }
            }
        });
    }
}
```

此时运行会打印出如图 12-11 所示的结果。

```
D/MainActivity: 登录成功了
D/MainActivity: response.code()==200
D/MainActivity: response.body().string()=={"result":"ok","data":{"userid":2,"username":"admin","userpwd":"123"}}
```

图 12-11

我们可以看到在 MainActivity 中，先获取登录按钮及用户名和密码输入框的实例，再给登录按钮注册一个单击事件，当单击"登录"按钮时调用 login()方法获取用户名和密码的值，通过 url 将数据发送到服务器端，服务器端将根据传送的用户名密码进行判断登录是否成功，并将结果返回给手机端。

注意事项：
- 异步请求不需要开启子线程，enqueue()方法会自动将网络请求部分放入子线程中执行。
- 回调接口的 onFailure()方法和 onResponse()方法执行在子线程。
- response.body().string()方法也必须放在子线程中。当执行这行代码得到结果后，再跳转到 UI 线程修改 UI。
- response.body().string()只能调用一次，在第一次时有返回值，第二次再调用时将会返回 null。

3. POST 的请求方式

POST 请求也分同步和异步两种方式，同步与异步的区别与 get()方法类似，所以此时只讲解 POST 异步请求的使用方法。

当我们使用 POST 方式时，需要通过 Request 的 post()方法传递参数，这些参数需要使用 FormBody 对象进行构造，然后再使用 Request 对象的 post()方法将参数传递出去。其他操作与在 post 请求中的操作一样。

```
private void postLogin() {
    //创建 okhttpClient 对象。
    okhttpClient client = new okhttpClient();
    //创建表单请求体
    RequestBody formBody = new FormEncodingBuilder()
                .add("username","admin")
                .add("password","123")
                .build();
    //创建 Request 对象
```

```
        Request request = new Request.Builder()
                .url("http://172.16.63.134:8080/apms/app/login")
                .post(formBody.build())//传递请求体
                .build();
           //创建 Call 对象
        Call call = client.newCall(request);
        //执行 POST 异步请求
call.enqueue(new Callback() {....});//此处省略回调方法
}
```

4. okhttp 上传文件

okhttp 进行文件上传实现与 Web 开发上传文件相同的方式,okhttp 封装了一个与 Web 一样的属性,即 multipart,用于上传文件,okhttp 也提供了上传文件的构造者 MultipartBuilder。只有这几个方法,使用键值对将要添加的信息传递进去。

那么怎么上传文件呢?接下来我们看一个上传文件的例子,步骤如下。

(1) 在 AndroidManifest.xml 文件中添加网络和读写 sdcard 的权限。

```xml
<uses-permission android:name="android.permission.INTERNET" />
<!-- 往 SDCard 写入数据权限 -->
<uses-permission android:name="android.permission.WRITE_EXTERNAL_STORAGE"/>
   <!-- 从 SDCard 读取数据权限 -->
<uses-permission android:name="android.permission.READ_EXTERNAL_STORAGE"/>
```

(2) 在 layout 文件夹添加 activity_main.xml。

```xml
<LinearLayout xmlns:android="http://schemas.android.com/apk/res/android"
    xmlns:app="http://schemas.android.com/apk/res-auto"
    xmlns:tools="http://schemas.android.com/tools"
    android:layout_width="match_parent"
    android:layout_height="match_parent"
    android:orientation="vertical"
    tools:context="com.hp.okhttp.MainActivity">

    <Button
        android:id="@+id/uplaod"
        android:layout_width="match_parent"
        android:layout_height="wrap_content"
        android:text="上传"
        android:layout_margin="10dip"/>
</LinearLayout>
```

(3) 在 MainActivity.java 中的具体引用与实现。

```java
import android.os.Bundle;
import android.os.Environment;
import android.support.annotation.Nullable;
import android.support.v7.app.AppCompatActivity;
import android.view.View;
```

```java
import android.widget.Button;

import com.hp.main.R;
import com.squareup.okhttp.Call;
import com.squareup.okhttp.Callback;
import com.squareup.okhttp.MediaType;
import com.squareup.okhttp.MultipartBuilder;
import com.squareup.okhttp.okhttpClient;
import com.squareup.okhttp.Request;
import com.squareup.okhttp.RequestBody;
import com.squareup.okhttp.Response;

import java.io.File;
import java.io.IOException;

public class    MainActivity    extends AppCompatActivity {
    private Button uplaod;
    @Override
    protected void onCreate(@Nullable Bundle savedInstanceState) {
        super.onCreate(savedInstanceState);
        setContentView(R.layout.activity_upload);
        uplaod = (Button)findViewById(R.id.uplaod);
        uplaod.setOnClickListener(new View.OnClickListener() {
            @Override
            public void onClick(View v) {
                doUpload();
            }
        });
    }
    public void doUpload() {
        //创建 okhttp 实例
        okhttpClient client = new okhttpClient();
        //获取上传图片的路径
        File file = new File(Environment.getExternalStorageDirectory(), "/launcher.png");
        if (!file.exists()) {
            return;
        }
        //创建 MultipartBuilder 实例
        MultipartBuilder multipartBuilder = new MultipartBuilder();
        //创建上传文件的 RequestBody 实例
        RequestBody fileBody = RequestBody.create(MediaType.parse("application/octet-stream"),file);
        RequestBody requestBody = multipartBuilder
                //设置请求类型为表单提交
                .type(MultipartBuilder.FORM)
                //设置上传的文件，第一参数为上传参数名称，第二个参数为图片的名称
                //第三个参数为上传文件的 RequestBody 对象
                .addFormDataPart("mfile", "launcher.png",fileBody)
```

```
                    .build();
            Request request = new Request.Builder()
                    .post(requestBody)
                    .url("http://172.16.63.134:8080/apms/app/upload")
                    .build();
        Call call= client.newCall(request);
          call.enqueue(new Callback() {
              @Override
              public void onFailure(Request request, IOException e) {

              }
              @Override
              public void onResponse(Response response) throws IOException {

              }
          });
    }
}
```

在 MainActivity 类中，upload()方法为上传图片的方法，在这个方法中，我们首先要创建 okhttp 的实例，然后获取图片路径的 File 对象，接着创建上传图片的 RequestBody 实例 fileBody，最后创建 MultipartBuilder 的实例，通过它的 type()方法设置请求类型。addFormDataPart()方法设置上传的 RequestBody 实例，再通过 build()方法创建上传请求的 RequestBody 的实例 requestBody。后面的与 post 请求一样直接请求网络即可。

5. okhttp 下载文件

okhttp 怎么下载文件呢？接下来我们看一个下载文件的例子，步骤如下。
(1) 在 AndroidManifest.xml 文件中添加网络和读写 sdcard 的权限。

```xml
<uses-permission android:name="android.permission.INTERNET" />
<!-- 往 SDCard 写入数据权限 -->
<uses-permission android:name="android.permission.WRITE_EXTERNAL_STORAGE"/>
 <!-- 从 SDCard 读取数据权限 -->
<uses-permission android:name="android.permission.READ_EXTERNAL_STORAGE"/>
```

(2) 在 layout 文件夹添加 activity_main.xml。

```xml
<LinearLayout xmlns:android="http://schemas.android.com/apk/res/android"
    xmlns:app="http://schemas.android.com/apk/res-auto"
    xmlns:tools="http://schemas.android.com/tools"
    android:layout_width="match_parent"
    android:layout_height="match_parent"
    android:orientation="vertical"
    tools:context="com.hp.okhttp.MainActivity">

    <Button
        android:id="@+id/download"
        android:layout_width="match_parent"
```

```xml
            android:layout_height="wrap_content"
            android:text="下载"
            android:layout_margin="10dip"/>
    <ImageView
        android:id="@+id/iv_result"
        android:layout_width="wrap_content"
        android:layout_height="wrap_content" />
</LinearLayout>
```

(3) 在 MainActivity.java 中的具体引用与实现。

```java
import android.graphics.Bitmap;
import android.graphics.BitmapFactory;
import android.os.Bundle;
import android.os.Environment;
import android.support.annotation.Nullable;
import android.support.v7.app.AppCompatActivity;
import android.view.View;
import android.widget.Button;
import android.widget.ImageView;

import com.hp.lpd.R;
import com.squareup.okhttp.Call;
import com.squareup.okhttp.Callback;
import com.squareup.okhttp.okhttpClient;
import com.squareup.okhttp.Request;
import com.squareup.okhttp.Response;

import java.io.File;
import java.io.FileOutputStream;
import java.io.IOException;
import java.io.InputStream;

public class MainActivity extends AppCompatActivity {
    private Button download;
    private ImageView iv_result;
    @Override
    protected void onCreate(@Nullable Bundle savedInstanceState) {
        super.onCreate(savedInstanceState);
        setContentView(R.layout.activity_download);
        download = (Button)findViewById(R.id.download);
        iv_result = (ImageView)findViewById(R.id.iv_result);
        download.setOnClickListener(new View.OnClickListener() {
            @Override
            public void onClick(View v) {
                downLoad();
            }
        });
```

```java
        }
        public void downLoad() {
            okhttpClient client = new okhttpClient();
            final Request request = new Request
                    .Builder()
                    .get()
                    .url("http://172.16.63.134:8080/apms/downdir/launcher.png")
                    .build();
            Call call = client.newCall(request);
            call.enqueue(new Callback() {
                @Override
                public void onFailure(Request request, IOException e) {

                }

                @Override
                public void onResponse(Response response) throws IOException {
                    InputStream inputStream = response.body().byteStream();
                    //将图片显示到 ImageView 中
                    final Bitmap bitmap = BitmapFactory.decodeStream(inputStream);
                    runOnUiThread(new Runnable() {
                        @Override
                        public void run() {
                            iv_result.setImageBitmap(bitmap);
                        }
                    });
                    //将图片保存到本地存储卡中
                    File file = new File(Environment.getExternalStorageDirectory(), "image.png");
                    FileOutputStream fileOutputStream = new FileOutputStream(file);
                    byte[] temp = new byte[128];
                    int length;
                    while ((length = inputStream.read(temp)) != -1) {
                        fileOutputStream.write(temp, 0, length);
                    }
                    fileOutputStream.flush();
                    fileOutputStream.close();
                    inputStream.close();
                }

            });
        }
    }
```

在 MainActivity 类中，download()方法为下载图片的方法，在该方法里，我们只要通过 okhttpClient 的 get()方法直接请求图片地址，然后在 onResponse()的方法里使用输入流获取图片的流，再转成图片在主线程将图片显示出来，将图片保存到本地存储卡中。效果如

图 12-12 所示。

【单元小结】

- 了解什么是网络编程。
- HTTP 和 URL 网络编程。
- WebView、Android 和 H5 混合编程。
- okhttp 网络框架。

【单元自测】

图 12-12

1. HttpURLConnection 是继承自( )的一个抽象类。
   A. URLConnection
   B. URL
   C. Connection
   D. UrlConnection
2. HttpClient 存在( )两种方式。
   A. GET 和 POST   B. doGet 和 doPost
   C. Server 和 Client   D. 提交和过滤
3. Android 中定义线程的方法与 Java 相同,可以使用两种方法:一种是( );另一种是( )。
   A. Thread 类   B. Runnable 接口
   C. run()方法   D. Thread 接口
4. okhttp 的异步请求方法是( )。
   A. execute()   B. enqueue()
   C. run()   D. exe()
5. Android 调用 js 代码的方法是( )。
   A. WebView 的 loadUrl()   B. WebView 的 reloadl()
   C. WebView 的 onPause()   D. WebView 的 clearHistory()

【上机实战】

上机目标

- 了解 Android 网络通信机制。
- 了解 okhttp 的用法。

上机练习

【问题描述】
编写服务器端程序，实现 Android 注册的接口，然后编写 Android 端注册功能的程序。

【问题分析】
- 在后台使用 JSP 后 SSH 编写服务器端的程序，实现用户注册功能的端口。
- 在 Android 编写注册界面，并通过 okhttp 的 POST 请求进行注册。

【参考步骤】
代码如下：

```java
// 服务器注册接口
import java.io.IOException;

import javax.servlet.http.HttpServletResponse;

import org.apache.struts2.ServletActionContext;

import com.opensymphony.xwork2.ActionSupport;
import com.sv.bean.Userinfo;
import com.sv.dao.UserinfoDAO;
import com.sv.util.GsonUtil;

public class UserRegisterAction extends ActionSupport{

    private static final long serialVersionUID = 1L;
    private UserinfoDAO userinfoDAO;
    private Userinfo userinfo;

    public void register(){
        try {
            userinfoDAO.save(userinfo);
            HttpServletResponse response = ServletActionContext.getResponse();
            response.setContentType("text/html;charset=utf-8");
            response.getWriter().write("{\"result\":\"ok\",\"data\":\"注册成功\"}");
        } catch (IOException e) {
            e.printStackTrace();
        }
    }

    public UserinfoDAO getUserinfoDAO() {
        return userinfoDAO;
    }

    public void setUserinfoDAO(UserinfoDAO userinfoDAO) {
        this.userinfoDAO = userinfoDAO;
```

```java
    }

    public Userinfo getUserinfo() {
        return userinfo;
    }

    public void setUserinfo(Userinfo userinfo) {
        this.userinfo = userinfo;
    }
}
```

// Android 端注册界面

```xml
<LinearLayout xmlns:android="http://schemas.android.com/apk/res/android"
    xmlns:app="http://schemas.android.com/apk/res-auto"
    xmlns:tools="http://schemas.android.com/tools"
    android:layout_width="match_parent"
    android:layout_height="match_parent"
    android:orientation="vertical"
    tools:context="com.hp.main.MainActivity">
    <LinearLayout
        android:layout_width="match_parent"
        android:layout_height="wrap_content"
        android:orientation="horizontal">
        <TextView
            android:layout_width="wrap_content"
            android:layout_height="wrap_content"
            android:text="用户名：" />
        <EditText
            android:id="@+id/username"
            android:layout_width="match_parent"
            android:layout_height="wrap_content" />
    </LinearLayout>
    <LinearLayout
        android:layout_width="match_parent"
        android:layout_height="wrap_content"
        android:orientation="horizontal">
        <TextView
            android:layout_width="wrap_content"
            android:layout_height="wrap_content"
            android:text="密码：" />
        <EditText
            android:id="@+id/password"
            android:layout_width="match_parent"
            android:layout_height="wrap_content" />
    </LinearLayout>
    <Button
```

```xml
            android:id="@+id/registerBt"
            android:layout_width="match_parent"
            android:layout_height="wrap_content"
            android:text="注册"
            android:layout_margin="10dip"/>
</LinearLayout>
```

```java
// Android 端注册 Activity 代码
import android.os.Bundle;
import android.support.v7.app.AppCompatActivity;
import android.util.Log;
import android.view.View;
import android.widget.Button;
import android.widget.EditText;

import com.hp.main.R;
import com.squareup.okhttp.Call;
import com.squareup.okhttp.Callback;
import com.squareup.okhttp.FormEncodingBuilder;
import com.squareup.okhttp.okhttpClient;
import com.squareup.okhttp.Request;
import com.squareup.okhttp.RequestBody;
import com.squareup.okhttp.Response;

import java.io.IOException;

public class MainActivity extends AppCompatActivity {
    private EditText username;
    private EditText password;
    private Button registerBt;
    @Override
    protected void onCreate(Bundle savedInstanceState) {
        super.onCreate(savedInstanceState);
        setContentView(R.layout.activity_main);
        username = (EditText)findViewById(R.id.username);
        password = (EditText)findViewById(R.id.password);
        registerBt = (Button)findViewById(R.id.registerBt);
        registerBt.setOnClickListener(new View.OnClickListener() {
            @Override
            public void onClick(View v) {
                login();
            }
        });
    }

    public void login(){
        String uname = username.getText().toString();
```

```java
            String pwd = password.getText().toString();
            String url = "http://172.16.63.134:8080/apms/app/regitser";
            okhttpClient client = new okhttpClient();
            //创建表单请求体
            RequestBody formBody = new FormEncodingBuilder()
                    .add("username",uname)
                    .add("password",pwd)
                    .build();
            Request request = new Request.Builder()
                    .url(url)
                    .patch(formBody)
                    .build();
            Call call = client.newCall(request);
            call.enqueue(new Callback() {
                @Override
                public void onFailure(Request request, IOException e) {

                }

                @Override
                public void onResponse(Response response) throws IOException {
                    if(response.isSuccessful()){
                        //回调的方法执行在子线程。
                        Log.d("MainActivity","注册成功了");
                        Log.d("MainActivity","response.code()=="+response.code());
                        Log.d("MainActivity","response.body().string()=="+response.body().string());
                    }
                }
            });
        }
    }
```

【拓展作业】

1. 举例说明 WebView 和 Android 原生的混合开发的方式。
2. 请编写程序实现简单的服务器和客户端网络编程功能。

单元十三

XML/JSON 数据解析

 课程目标

- XML 数据解析
- JSON 数据解析

简介

在 Android 系统中，从后台服务器上获取的数据往往是 XML 或 JSON 格式，本单元针对 XML 和 JSON 数据解析的内容做一个简单介绍。

13.1 XML 数据解析

在通常情况下，每个需要访问网络的应用程序都会有一个自己的服务器，我们不仅可以向服务器提交数据，也可以从服务器上获取数据。不过这个过程中就出现了一个问题，这些数据要以什么格式在网络上传输呢？随便传递一段文本肯定是不行的，因为另一方根本就不会知道这段文本的用途是什么。因此，一般我们都会在网络上传输一些格式化后的数据，这种数据会有一定的结构规格和语义，当另一方收到数据消息后就可以按照相同的结构规格进行解析，从而获取想要的那部分内容。

在网络上传输数据时最常用的格式有两种，即 XML 和 JSON，下面我们就逐个地进行学习，本节首先学一下如何解析 XML 格式的数据。解析 XML 格式的数据有多种方式，本节我们学习比较常用的两种，Pull 解析和 DOM 解析。

13.1.1 Pull 解析方式

Pull 解析器的运行方式是基于事件的模式。在 Pull 解析过程中返回的是数字，而且我们需要自己获取产生的事件然后做相应的操作。

简单起见，这里需要一个 XML 格式的数据，这样我们就可将数据放在 assets 文件夹里，从而把工作的重心放在 XML 数据解析上。

Assets 下的 students.xml 文件的代码如下所示。

```xml
<?xml version="1.0" encoding="utf-8"?>
<students>
    <student>
        <name>张三</name>
        <sex>男</sex>
        <age>19</age>
    </student>
    <student>
        <name>李四</name>
        <sex>女</sex>
        <age>18</age>
    </student>
    <student>
        <name>王五</name>
        <sex>男</sex>
        <age>20</age>
```

```
        </student>
    </students>
```

完成 XML 格式的数据后,现在从中解析出我们想要得到的那部分内容。修改 MainActivity.java 中的代码,如下所示。

```
import android.os.Bundle;
import android.support.v7.app.AppCompatActivity;
import android.util.Log;

import com.hp.main.R;

import org.xmlpull.v1.XmlPullParser;
import org.xmlpull.v1.XmlPullParserFactory;

import java.io.InputStream;

public class MainActivity extends AppCompatActivity{

    @Override
    protected void onCreate(Bundle savedInstanceState) {
        super.onCreate(savedInstanceState);
        setContentView(R.layout.activity_main);
        parseXMLWithPull();
    }
    //用 Pull 方式解析 XML
    private   void parseXMLWithPull(){
        try {
            //获取 xml 文档的输入流
            InputStream in = getAssets().open("student.xml");
            //创建 XmlPullParserFactory 实例
            XmlPullParserFactory factory = XmlPullParserFactory.newInstance();
            //创建 XmlPullParser 实例
            XmlPullParser xmlPullParser = factory.newPullParser();
            //将 Xml 的输入流加入解析器
            xmlPullParser.setInput(in,"utf-8");
            //获取当前解析事件,返回的是数字
            int eventType = xmlPullParser.getEventType();
            //保存解析的内容
            String name = "";
            String sex = "";
            int age = 0;
            //判断是否是文档的结束标签
            while (eventType != (XmlPullParser.END_DOCUMENT)){
                String nodeName = xmlPullParser.getName();
                switch (eventType){
                    //开始解析 XML
                    case XmlPullParser.START_TAG:{
```

```java
                        //nextText()用于获取节点内的具体内容
                        if("name".equals(nodeName))
                            name = xmlPullParser.nextText();
                        else if("sex".equals(nodeName))
                            sex = xmlPullParser.nextText();
                        else if("age".equals(nodeName))
                            age = Integer.parseInt(xmlPullParser.nextText());
                    }
                    break;
                    //结束解析
                    case XmlPullParser.END_TAG:{
                        if("student".equals(nodeName)){
                            Log.d("MainActivity", "parseXMLWithPull: 姓名:    "+ name);
                            Log.d("MainActivity", "parseXMLWithPull: 性别:    "+ sex);
                            Log.d("MainActivity", "parseXMLWithPull: 年龄:    "+ age);
                        }
                    }
                    break;
                    default:
                        break;
                }
                //获取下一个标签
                eventType = xmlPullParser.next();
            }
        } catch (Exception e) {
            e.printStackTrace();
        }

    }
}
```

下面我们仔细看一下parseXMLWithPull()方法中的代码。这里首先要获取XML文档的输入流,再获取到一个XmlPullParserFactor的实例,并借助这个实例得到 XmlPullParser 对象,然后调用XmlPullParser的setInput()方法将获取的XML数据设置进去即可开始解析。解析的过程也非常简单,通过getEventType()方法可以得到当前的解析事件,然后在一个while循环中不断地进行解析,如果当前的解析事件不等于XmlPullParser.END_DOCUMENT,则说明解析工作还没完成,调用 next()方法后可以获取下一个解析事件。

在 while 循环中,我们通过 getName()方法得到当前节点的名字,如果发现节点名等于name、sex 或 age,就调用 nextText()方法来获取节点内具体的内容,每当解析完一个 student 节点后就将获取到的内容打印出来。下面我们测试一下,运行PullXmlTest项目,观察LogCat中的打印日志,结果如图 13-1 所示。

```
D/MainActivity: parseXMLWithPull: 姓名：张三
D/MainActivity: parseXMLWithPull: 性别：男
D/MainActivity: parseXMLWithPull: 年龄：19
D/MainActivity: parseXMLWithPull: 姓名：李四
D/MainActivity: parseXMLWithPull: 性别：女
D/MainActivity: parseXMLWithPull: 年龄：18
D/MainActivity: parseXMLWithPull: 姓名：王五
D/MainActivity: parseXMLWithPull: 性别：男
D/MainActivity: parseXMLWithPull: 年龄：20
```

图 13-1

13.1.2 DOM 解析方式

解析 DOM，即对象文档模型，它是将整个 XML 文档载入内存(所以效率较低，不推荐使用)，每一个节点当作一个对象，结合代码分析。DOM 实现时首先为 XML 文档的解析定义一组接口，解析器读入整个文档，然后构造一个驻留内存的树结构，这样代码就可以使用 DOM 接口来操作整个树结构。由于 DOM 在内存中以树形结构存放，因此检索和更新效率会更高。但是对于特别大的文档，解析和加载整个文档将会很耗资源。当然，如果 XML 文件的内容比较小，采用 DOM 是可行的。

使用 DOM 对 XML 文件进行操作时，首先要解析文件，将文件分为独立的元素、属性和注释等，然后以节点树的形式在内存中对 XML 文件进行表示，就可以通过节点树访问文档的内容，并根据需要修改文档。

assets 下的 students.xml 文件的代码如下所示。

```xml
<?xml version="1.0" encoding="utf-8"?>
<students>
    <student>
        <name>张三</name>
        <sex>男</sex>
        <age>19</age>
    </student>
    <student>
        <name>李四</name>
        <sex>女</sex>
        <age>18</age>
    </student>
    <student>
        <name>王五</name>
        <sex>男</sex>
        <age>20</age>
    </student>
</students>
```

既然 XML 格式的数据已经提供好了，现在要做的就是从中解析出我们想要得到的那部分内容。修改 MainActivity.java 中的代码，如下所示。

```java
import android.os.Bundle;
import android.support.v7.app.AppCompatActivity;
import android.util.Log;

import com.hp.main.R;

import org.w3c.dom.Document;
import org.w3c.dom.Node;
import org.w3c.dom.NodeList;

import java.io.InputStream;

import javax.xml.parsers.DocumentBuilder;
import javax.xml.parsers.DocumentBuilderFactory;

public class MainActivity extends AppCompatActivity{

    @Override
    protected void onCreate(Bundle savedInstanceState) {
        super.onCreate(savedInstanceState);
        setContentView(R.layout.activity_main);
        parseXMLWithDom();
    }
    //用 Pull 方式解析 XML
    private  void parseXMLWithDom(){
        //获取 xml 文档的输入流
        try {
            InputStream in = getAssets().open("student.xml");
            DocumentBuilderFactory factory = DocumentBuilderFactory.newInstance();
            DocumentBuilder builder = factory.newDocumentBuilder();
            //获得 Document 对象
            Document document = builder.parse(in);
            //获得 student 的 List
            NodeList studentList = document.getElementsByTagName("student");
            //遍历 student 标签
            for (int i = 0; i < studentList.getLength(); i++) {
                //获得 student 标签
                Node node_student = studentList.item(i);
                //获得 student 标签里的标签
                NodeList childNodes = node_student.getChildNodes();
                //遍历 student 标签里的标签
                for (int j = 0; j < childNodes.getLength(); j++) {
                    //获得 name 和 nickName 标签
                    Node childNode = childNodes.item(j);
                    //判断是 name 还是 nickName
                    if ("name".equals(childNode.getNodeName())) {
                        String name = childNode.getTextContent();
```

```
                    Log.d("MainActivity", "parseXMLWithDom: 姓名：  "+ name);
                } else if ("sex".equals(childNode.getNodeName())) {
                    String sex = childNode.getTextContent();
                    Log.d("MainActivity", "parseXMLWithDom: 性别：  "+ sex);
                }else if ("age".equals(childNode.getNodeName())) {
                    String age = childNode.getTextContent();
                    Log.d("MainActivity", "parseXMLWithDom: 年龄：  "+ age);
                }
            }
        }
    } catch (Exception e) {
        e.printStackTrace();
    }
  }
}
```

下面仔细看一下 parseXMLWithDom()方法中的代码。这里首先要获取 XML 文档的输入流，然后获取到一个 DocumentBuilderFactory 的实例，并借助这个实例得到 DocumentBuilder 对象，然后调用 DocumentBuilder 的 parse()方法将获取的 XML 数据设置进去即可开始解析。解析的过程也非常简单，通过 getElementsByTagName()方法可以得到需要解析的标签列表，然后在一个 for 循环中不断地进行解析，获取当前标签的所有子标签，再判断子标签是否为需要的标签，通过 getNodeName()方法可以获取当前标签的内容。

下面我们测试一下，运行 PullXmlTest 项目，观察 LogCat 中的打印日志，结果如图 13-2 所示。

```
D/MainActivity: parseXMLWithDom: 姓名： 张三
D/MainActivity: parseXMLWithDom: 性别： 男
D/MainActivity: parseXMLWithDom: 年龄： 19
D/MainActivity: parseXMLWithDom: 姓名： 李四
D/MainActivity: parseXMLWithDom: 性别： 女
D/MainActivity: parseXMLWithDom: 年龄： 18
D/MainActivity: parseXMLWithDom: 姓名： 王五
D/MainActivity: parseXMLWithDom: 性别： 男
D/MainActivity: parseXMLWithDom: 年龄： 20
```

图 13-2

可以看到，我们已经将 XML 数据中的指定内容成功解析出来了。

13.2 JSON 数据解析

JSON(JavaScript Object Notation)是一种轻量级的数据交换格式，它是基于 JavaScript(Standard ECMA-262 3rd Edition - December 1999)的一个子集。JSON 采用完全独立于语言的文本格式，但是也使用了类似于 C 语言家族的习惯(包括 C、C++、C#、Java、JavaScript、Perl、Python 等)。这些特性使 JSON 成为理想的数据交换语言，易于人们阅读和编写，同时也易于机器解析和生成。

13.2.1 基础结构

简单来说，JSON 是 JavaScript 中的对象和数组，所以 JSON 有两种结构，即对象和数组，通过这两种结构可以表示各种复杂的结构。

1. 对象

对象在 JSON 中表示为"{}"括起来的内容，数据结构为{key：value,key：value,...}的键值对的结构。在面向对象的语言中，key 为对象的属性，value 为对应的属性值，所以取值方法为"对象.key"，以此来获取属性值，属性值的类型可以是数字、字符串、数组及对象等。

2. 数组

数组在 JSON 中是中括号"[]"括起来的内容，数据结构为 ["java","javascript","vb",...]，取值方式与所有语言中一样，使用索引获取，字段值的类型可以是数字、字符串、数组、对象几种。

经过对象、数组两种结构即可组合成复杂的数据结构。

13.2.2 JSON 数据解析

在实际的开发中，通常使用 JSON 作为客户端与服务端的数据交换格式，所以我们必须了解 JSON 数据的解析方法，这里介绍两种解析 JSON 数据的方法，分别是 Android 系统自带的 Android 原生技术和 Google 提供的 Gson 解析。

1. 使用 Android 原生技术

Android 的 JSON 解析部分都在包 org.json 下，主要有以下两个类。

1) JSONObject 类

(1) 可以看作一个 JSON 对象，这是系统中有关 JSON 定义的基本单元，其包含一对(Key/Value)数值。

(2) 它对外部引用 toString()方法输出的数值，对内部引用 put()方法添加数值，如 new JSONObject().put("JSON","Hello, World!")。

(3) Value 的类型包括 Boolean、JSONArray、JSONObject、Number、String，或者默认值 JSONObject.NULL object。

2) JSONArray 类

(1) 它代表一组有序的数值。

(2) get()和 opt()两种方法都可以通过 index 索引返回指定的数值。

(3) put()方法用来添加或者替换数值。

(4) Value 类型包括 Boolean、JSONArray、JSONObject、Number、String，或者默认值 JSONObject.NULL object。

使用 JSONObject、JSONArray 构建 json 文本，代码如下。

```json
{
    "name": "张三",
    "sex": "男",
    "age": "22",
    "score":
    {
        "math": "60",
        "chinese": "61",
        "english": "62",
        "compre":
        [
            {
                "arts": "100"
            },
            {
                "science": "110"
            }
        ]
    }
}
```

修改 MainActivity.java 的代码，如下所示。

```java
import android.os.Bundle;
import android.support.annotation.Nullable;
import android.support.v7.app.AppCompatActivity;
import android.util.Log;

import com.hp.main.R;

import org.json.JSONArray;
import org.json.JSONObject;

import java.io.InputStream;

public class MainActivity extends AppCompatActivity {

    @Override
    protected void onCreate(@Nullable Bundle savedInstanceState) {
        super.onCreate(savedInstanceState);
        setContentView(R.layout.activity_main);
        try {
            InputStream in = getAssets().open("json.txt");
            byte[] bytes = new byte[in.available()];
            in.read(bytes);
            String json = new String(bytes,"UTF-8");
            Log.d("MainActivity",json);
```

```java
            //创建 JSONObject 对象
            JSONObject jsonObject = new JSONObject(json);
            //获取学生姓名、性别、年龄
            String name = jsonObject.getString("name");
            Log.d("MainActivity","学生姓名: "+name);
            String sex = jsonObject.getString("sex");
            Log.d("MainActivity","学生性别: "+sex);
            String age = jsonObject.getString("age");
            Log.d("MainActivity","学生年龄: "+age);
            //获取学生学分，这里学分是{}格式，那就要用 JSONObject 接受，然后再解析
            JSONObject scoreObject = jsonObject.getJSONObject("score");
            //获取学生数学、语文、英文成绩
            String math = scoreObject.getString("math");
            Log.d("MainActivity","数学成绩: "+math);
            String chinese = scoreObject.getString("chinese");
            Log.d("MainActivity","语文成绩: "+chinese);
            String english = scoreObject.getString("english");
            Log.d("MainActivity","英文成绩: "+english);
            //综合成绩是[]格式，那就需要用 JSONArray 接受
            JSONArray jsonArray = scoreObject.getJSONArray("compre");
            JSONObject artsObject = (JSONObject)jsonArray.get(0);
            Log.d("MainActivity","文科综合成绩: "+artsObject.getString("arts"));
            JSONObject scienceObject = (JSONObject)jsonArray.get(1);
            Log.d("MainActivity","理科综合成绩: "+scienceObject.getString("science"));
        } catch (Exception e) {
            e.printStackTrace();
        }
    }
}
```

首先我们需要将 asset 下的 json.txt 文本中的 JSON 数据读取处理，然后通过 JSONObject 对象的构造器将 JSON 字符串转成 JSONObject 对象。可以看到，学生的姓名、性别、年龄可以直接解析，而学生分数定义成 JSON 数组，所以我们要先获取分数数组并传入一个 JSONArray 对象中，然后再解析 JSONArray 数组，从中取出的每一个元素都是一个 JSONObject 对象，接下来只需要调用 getString()方法将这些数据取出并打印出来即可，结果如图 13-3 所示。

```
D/MainActivity: {"name":"张三","sex":"男","age":"22","score":{"math":"60","chinese":"61","english":"62","compre":[{"arts":"100"},{"science":"110"}]}}
D/MainActivity: 学生姓名: 张三
D/MainActivity: 学生性别: 男
D/MainActivity: 学生年龄: 22
D/MainActivity: 数学成绩: 60
D/MainActivity: 语文成绩: 61
D/MainActivity: 英文成绩: 62
D/MainActivity: 文科综合成绩: 100
D/MainActivity: 理科综合成绩: 110
```

图 13-3

2. 使用 GSON

如果我们认为使用 JSONObject 来解析 JSON 数据已经非常简单了,那就太容易满足了。谷歌提供的 GSON 开源库可以让解析 JSON 数据的工作更简单。但是 GSON 并没有被添加到 Android 官方的 API 中,所以如果想要使用这个功能,则必须在项目中添加一个 GSON 的 Jar 包,将 GSON 的 jar 包拷贝到项目的 libs 目录下,GSON 库就会自动添加到项目中,如图 13-4 所示。

图 13-4

那么 GSON 库究竟有哪些功能呢?其实它主要就是可以将一段 JSON 格式的字符串自动映射成一个对象,从而不需要我们再手动编写代码进行解析。

例如,下面一段 JSON 格式的数据。

{"name":"张三","sex":"男","age":"22"}

因为 GSON 是将字符串直接映射成对象,所以我们可以定义一个 Student 类,并加入 name、sex 和 age 3 个字段,Student 对象的代码如下所示。

```
public class Student {
    private String name;
    private String sex;
    private int age;

    public String getName() {
        return name;
    }

    public void setName(String name) {
        this.name = name;
    }

    public String getSex() {
        return sex;
    }

    public void setSex(String sex) {
        this.sex = sex;
    }

    public int getAge() {
        return age;
```

```java
    }

    public void setAge(int age) {
        this.age = age;
    }
}
```

然后只需简单地调用如下代码就可以将 JSON 数据自动解析成一个 Student 对象，MainActivity.java 的代码如下所示。

```java
import android.os.Bundle;
import android.support.annotation.Nullable;
import android.support.v7.app.AppCompatActivity;
import android.util.Log;

import com.google.gson.Gson;
import com.hp.main.R;
import com.hp.model.Student;

import java.io.InputStream;

public class MainActivity extends AppCompatActivity {

    @Override
    protected void onCreate(@Nullable Bundle savedInstanceState) {
        super.onCreate(savedInstanceState);
        setContentView(R.layout.activity_main);
        try {
            InputStream in = getAssets().open("student.txt");
            byte[] bytes = new byte[in.available()];
            in.read(bytes);
            String json = new String(bytes,"UTF-8");
            Log.d("MainActivity",json);
            Gson gson = new Gson();
            Student student = gson.fromJson(json, Student.class);
            Log.d("MainActivity","学生姓名： "+student.getName());
            Log.d("MainActivity","学生性别： "+student.getSex());
            Log.d("MainActivity","学生年龄： "+student.getAge());
        } catch (Exception e) {
            e.printStackTrace();
        }
    }
}
```

现在重新运行一下程序，结果如图 13-5 所示。

```
D/MainActivity: {"name":"张三","sex":"男","age":"22"}
D/MainActivity: 学生姓名：张三
D/MainActivity: 学生性别：男
D/MainActivity: 学生年龄：22
```

图 13-5

我们知道 JSON 数据还有数组格式，那么 JSON 数组怎么解析呢？例如，下面一段 JSON 格式的数据。

[{"name":"张三","sex":"男","age":"22"},{"name":"李四","sex":"女","age":"20"},{"name":"王五","sex":"男","age":"28"}]

如果需要解析的是一段 JSON 数组，那么会稍微麻烦一点儿，我们需要借助 TypeToken 将期望解析成的数据类型传入 fromJson()方法中，MainActivity.java 代码如下所示。

```java
import com.google.gson.Gson;
import com.google.gson.reflect.TypeToken;
import com.hp.main.R;
import com.hp.model.Student;

import java.io.InputStream;
import java.util.List;

public class MainActivity extends AppCompatActivity {

    @Override
    protected void onCreate(@Nullable Bundle savedInstanceState) {
        super.onCreate(savedInstanceState);
        setContentView(R.layout.activity_main);
        try {
            InputStream in = getAssets().open("students.txt");
            byte[] bytes = new byte[in.available()];
            in.read(bytes);
            String json = new String(bytes,"UTF-8");
            Log.d("MainActivity",json);
            Gson gson = new Gson();
            List<Student> students = gson.fromJson(json, new TypeToken<List<Student>>()
            {}.getType());
            for (Student student : students) {
                Log.d("MainActivity", "学生姓名：" + student.getName());
                Log.d("MainActivity", "学生性别：" + student.getSex());
                Log.d("MainActivity", "学生年龄：" + student.getAge());
            }
        } catch (Exception e) {
            e.printStackTrace();
        }
    }
}
```

现在重新运行一下程序，结果如图13-6所示。

```
D/MainActivity: [{"name":"张三","sex":"男","age":"22"},{"name":"李四","sex":"女","age":"20"},{"name":"王五","sex":"男","age":"28"}]
D/MainActivity: 学生姓名：张三
D/MainActivity: 学生性别：男
D/MainActivity: 学生年龄：22
D/MainActivity: 学生姓名：李四
D/MainActivity: 学生性别：女
D/MainActivity: 学生年龄：20
D/MainActivity: 学生姓名：王五
D/MainActivity: 学生性别：男
D/MainActivity: 学生年龄：28
```

图 13-6

3. 使用 GSON 解析网络 JSON 数据

我们的手机往往需要网络通信，那么在获取到服务器响应的数据后就需要对它进行解析和处理。需要注意的是，网络请求通常返回的都是 JSON 格式的数据，使用时需要进行解析才可以展示到 UI 界面。例如，我们获取后台返回的用户列表的数据如图 13-7 所示。

```
D/MainActivity: [{"userid":1,"username":"tom","userpwd":"123"},{"userid":2,"username":"jack","userpwd":"123"},{"userid":3,"username":"mark","userpwd":"123"}]
```

图 13-7

MainActivity.java：

```java
import android.os.Bundle;
import android.support.annotation.Nullable;
import android.support.v7.app.AppCompatActivity;
import android.util.Log;

import com.google.gson.Gson;
import com.google.gson.reflect.TypeToken;
import com.hp.lpd.R;
import com.hp.model.Userinfo;
import com.squareup.okhttp.Call;
import com.squareup.okhttp.Callback;
import com.squareup.okhttp.OkHttpClient;
import com.squareup.okhttp.Request;
import com.squareup.okhttp.Response;

import java.io.IOException;
import java.util.List;

public class GsonActivity1 extends AppCompatActivity {

    @Override
    protected void onCreate(@Nullable Bundle savedInstanceState) {
        super.onCreate(savedInstanceState);
        setContentView(R.layout.activity_pull);
        list();
```

```java
        }

    public void list(){
        String url = "http://192.168.0.102:8080/apms/app/list";
        OkHttpClient client = new OkHttpClient();
        Request request = new Request.Builder()
                .url(url)
                .build();
        Call call = client.newCall(request);
        call.enqueue(new Callback() {
            @Override
            public void onFailure(Request request, IOException e) {

            }

            @Override
            public void onResponse(Response response) throws IOException {
                if(response.isSuccessful()){
                    //回调的方法执行在子线程
                    String result = response.body().string();
                    Log.d("MainActivity",result);
                    Gson gson = new Gson();
                    List<Userinfo> userinfoList = gson.fromJson(result, new TypeToken<List<Userinfo>>()
                    {}.getType());
                    for (Userinfo userinfo : userinfoList) {
                        Log.d("MainActivity", "用户 ID： " + userinfo.getUserid());
                        Log.d("MainActivity", "用户姓名： " + userinfo.getUsername());
                        Log.d("MainActivity", "用户密码： " + userinfo.getUserpwd());
                    }
                }
            }
        });
    }
}
```

现在重新运行一下程序，结果如图 13-8 所示。

```
D/MainActivity: 用户ID: 1
D/MainActivity: 用户姓名: tom
D/MainActivity: 用户密码: 123
D/MainActivity: 用户ID: 2
D/MainActivity: 用户姓名: jack
D/MainActivity: 用户密码: 123
D/MainActivity: 用户ID: 3
D/MainActivity: 用户姓名: mark
D/MainActivity: 用户密码: 123
```

图 13-8

【单元小结】

- 了解 JSON 数据的概念。
- 理解 Android 的 xml 解析。
- 理解 Android 的 json 数据解析。

【单元自测】

1. 网络上常见的传输数据的格式是(　　)。
   A. xml 和 json　　　　　　　　　B. json 和 html
   C. html 和 xml　　　　　　　　　D. txt 和 html
2. pull 解析的 XmlPullParser 接口中只需要调用(　　)方法就可以获取下一个事件类型。
   A. nextText()　　　　　　　　　　B. next()
   C. nextContext()　　　　　　　　D. getTextContent()
3. JSON 数据有(　　)两种格式。
   A. List 和数组　　　　　　　　　B. 对象和 List
   C. Map 和 List　　　　　　　　　D. 对象和数组
4. Android 中 GSON 将 JSON 数据解析成对象的方法是(　　)。
   A. toJson()　　　　　　　　　　　B. toObject()
   C. fromJson()　　　　　　　　　　D. fromObject()
5. GSON 在解析 JSON 数组时使用的方法是(　　)。
   A. GsonToken()　　　　　　　　　B. TypeToken()
   C. TypeGson()　　　　　　　　　　D. TypeList()

【上机实战】

上机目标

- 了解 Android 网络通信机制。
- 了解客户端获取服务器端数据并解析的方法。

上机练习

【问题描述】
编写服务器端程序,返回用户列表,然后编写客户端获取数据并解析。

单元十三 XML/JSON数据解析

【问题分析】
- 在 Activity 中编写客户器端的程序，实现 JSON 数据的获取和解析。
- 在服务端中实现客户端请求接口。

【参考步骤】
代码如下：

```java
//服务器端
import org.apache.struts2.ServletActionContext;

import com.opensymphony.xwork2.ActionSupport;
import com.sv.bean.Userinfo;
import com.sv.dao.UserinfoDAO;
import com.sv.util.GsonUtil;

public class UserinfoAction extends ActionSupport{

    private static final long serialVersionUID = 1L;
    private UserinfoDAO userinfoDAO;

    public void list(){
        HttpServletResponse response = ServletActionContext.getResponse();
        response.setContentType("text/html;charset=utf-8");
        try {
            List list = userinfoDAO.list();
            String jsonString = GsonUtil.GsonString(list);
            response.getWriter().write(jsonString);
        } catch (IOException e) {
            e.printStackTrace();
        }
    }
    public UserinfoDAO getUserinfoDAO() {
        return userinfoDAO;
    }

    public void setUserinfoDAO(UserinfoDAO userinfoDAO) {
        this.userinfoDAO = userinfoDAO;
    }

}

// 客户端
import android.os.Bundle;
import android.support.annotation.Nullable;
import android.support.v7.app.AppCompatActivity;
import android.util.Log;
```

```java
import com.google.gson.Gson;
import com.google.gson.reflect.TypeToken;
import com.hp.lpd.R;
import com.hp.model.Userinfo;
import com.squareup.okhttp.Call;
import com.squareup.okhttp.Callback;
import com.squareup.okhttp.OkHttpClient;
import com.squareup.okhttp.Request;
import com.squareup.okhttp.Response;

import java.io.IOException;
import java.util.List;

public class GsonActivity1 extends AppCompatActivity {

    @Override
    protected void onCreate(@Nullable Bundle savedInstanceState) {
        super.onCreate(savedInstanceState);
        setContentView(R.layout.activity_pull);
        list();
    }

    public void list(){

        String url = "http://192.168.0.102:8080/apms/app/list";
        OkHttpClient client = new OkHttpClient();
        Request request = new Request.Builder()
                .url(url)
                .build();
        Call call = client.newCall(request);
        call.enqueue(new Callback() {
            @Override
            public void onFailure(Request request, IOException e) {

            }

            @Override
            public void onResponse(Response response) throws IOException {
                if(response.isSuccessful()){
                    //回调的方法执行在子线程
                    String result = response.body().string();
                    Log.d("MainActivity",result);
                    Gson gson = new Gson();
                    List<Userinfo> userinfoList = gson.fromJson(result, new
                            TypeToken<List<Userinfo>>()
                    {}.getType());
                    for (Userinfo userinfo : userinfoList) {
```

```
                    Log.d("MainActivity", "用户 ID: " + userinfo.getUserid());
                    Log.d("MainActivity", "用户姓名: " + userinfo.getUsername());
                    Log.d("MainActivity", "用户密码: " + userinfo.getUserpwd());
                }
            }
        });
    }
}
```

【拓展作业】

1. 举例说明 Android 获取数据的格式有哪些？
2. 请编写程序实现简单的服务器和客户端通信并解析数据。